To Gregg Bemi[illegible]

independent investigator,

with gratitude for all your

help on this project.

Peter R. Limburg

5/5/2004

Deep-Sea Detectives

Deep-Sea Detectives

MARITIME MYSTERIES AND FORENSIC SCIENCE

Peter R. Limburg

ECW PRESS

Published by ECW PRESS
2120 Queen Street East, Suite 200, Toronto, Ontario, Canada M4E 1E2

NATIONAL LIBRARY OF CANADA CATALOGUING IN PUBLICATION DATA

Limburg, Peter R.
Deep-sea detectives : maritime mysteries and forensic science / Peter R. Limburg.
ISBN 1-55022-578-2
1. Marine accidents—Investigation. 2. Deep diving. I. Title.

G525.154 2003 363.12'3 C2003-902187-4

Acquisition and production: Emma McKay
Copy editor: Jodi Lewchuk
Design and typesetting: Yolande Martel
Cover design: Paul Hodgson
Printing: Quebecor

Diligent efforts have been made to contact copyright holders; please excuse any inadvertent errors or omissions, but if anyone has been unintentionally omitted, the publisher would be pleased to receive notification and make acknowledgements in future printings.

This book is set in Utopia and Trade Gothic

The publication of *Deep-Sea Detectives* has been generously supported by the Canada Council, by the Government of Ontario through the Ontario Media Development Corporation's Ontario Book Initiative, by the Ontario Arts Council, and by the Government of Canada through the Book Publishing Industry Development Program. Canadä

DISTRIBUTION

CANADA: Jaguar Book Group, 100 Armstrong Avenue, Georgetown, Ontario L7G 5S4

UNITED STATES: Independent Publishers Group, 814 North Franklin Street, Chicago, Illinois 60610

EUROPE: Turnaround Publisher Services, Unit 3, Olympia Trading Estate, Coburg Road, Wood Green, London N2Z 6T2

AUSTRALIA AND NEW ZEALAND: Wakefield Press, 1 The Parade West (Box 2266), Kent Town, South Australia 5071

PRINTED AND BOUND IN CANADA

ECW PRESS
ecwpress.com

To my wife, Maggie, as always;
To the victims of the tragedies described in this book;
And to all the people who, in one way or another, helped me to put it together.

CONTENTS

FOREWORD

The idea for this book came to me as I was working on an article for *Science Year*, formerly a longtime client. The article was on deep-water archaeology, and my research led me to the web site of Texas A & M University, a leader in marine archaeology. While fumbling my way around that site (I am not an expert navigator of the web), I stumbled upon a link to the story of the *Lucona*, a Panamanian-registered freighter, allegedly carrying a valuable uranium-ore processing mill, that exploded and sank under suspicious circumstances in the Indian Ocean. Although the wreck lay at a depth of nearly 15,000 feet, it was located and identified through the use of up-to-date underwater technology: side-scan sonar and ROVs (the so-called *robot submarines*).

The cargo belonged to a shady Austrian businessman, who had insured it for a very high sum. The insurance company, delighted to get the premium for insuring the cargo, suspected fraud and refused to pay. The case eventually grew into a major political scandal and brought down the then current government of Austria. I thought it was a nifty story—but then I realized that my readership would be almost entirely in the English-speaking world, and few people in the U.S., Canada, the U.K., or Australia would be interested in a book on an Austrian shipwreck-cum-scandal. So the concept expanded into a story on deep-sea detection and the conclusions that experts draw from the mute evidence of a wrecked ship or aircraft.

I decided against covering the *Titanic* (about which so much has been written) or marine archaeology (fascinating as it is) because marine archaeologists are not primarily interested in what has caused a ship to sink; rather, they are concerned with the contents of its cargo and the details of its construction, if enough of the hull survives to give any clues. And that was not my focus.

It is said that a journey of a thousand miles starts with a single step. My first step was to talk with the son of friends, a bright and personable young man who publishes a sort of maritime/financial newsletter. A partner of his put me in touch with the New York maritime representative of the Republic of Vanuatu, who in turn told me about SNAME (pronounced "snay-mee"). "What's SNAME?" I made bold to ask. He explained that it was the Society of Naval Architects and Marine Engineers, and he also gave me the names of a few other experts to contact. With that, I was off and running.

Researching this book gave me an education not only in the technology of ROVs, side-scan sonar, and other electronic/mechanical marvels of our age but also in such interesting arcana as *green water,* which turned out to be a solid wall of water crashing across a ship's deck, as distinct from individual waves or spray. Green water is often mentioned in maritime accident reports, and it is a thing to be dreaded by seafarers. (By the way, it is usually gray rather than green.)

I hope that the readers of this book will find the stories covered herein as fascinating as I did and will respect the memories of the crew members and officers who died in these tragedies.

Peter R. Limburg
February 2004

PROLOGUE

As night fell over the Havana harbor on February 15, 1898, the USS *Maine,* a powerful battleship, bobbed quietly in the water next to a Spanish warship. Having docked a month earlier, the *Maine*'s mission was to show her flag and protect Americans and American enterprises in Spanish-occupied Cuba.

This was no easy undertaking given the simmering tensions on the island; despite the best efforts of the stern and brutal governor, Spanish General Valeriano Weyler, armed Cuban rebels threatened to rise up. Most people in the United States sympathized with these rebels, and rival newspaper owners William Randolph Hearst and Joseph Pulitzer competed for readers with lurid stories of Spanish atrocities against the Cuban people. Meanwhile, powerful American interests envious of Europe's empires (Britain, Germany, France, the Netherlands, and even insignificant Belgium all had profitable foreign holdings) believed the United States should join the ranks of the world's Great Powers—and Spain's Caribbean islands lay close at hand, a tempting prize. Humanitarians and imperialists both pressed for American intervention in Cuba.

President McKinley responded by dispatching the *Maine* to Havana. But on this February night, the ship's tenure in the harbor ended abruptly. At 9:40 P.M., a thunderous explosion ripped the *Maine* apart, sinking it almost instantly and killing 266 of the 350 crew.

The first investigation took place shortly after the sinking; both the Spanish and the U.S. navies sent divers to the bottom of the harbor to examine the wreckage. Not surprisingly, each team came back with a different report. The Americans claimed the inward-bent and twisted hull plates were evidence of an external explosion—a detonated mine—that set off ammunition stored in the ship's magazines. The Spaniards, whose government very much wanted to avoid war with the United States, found no indication of foul play. Nonetheless, the sinking of the *Maine* furnished a convenient pretext for America to declare war on Spain just months later (and ultimately take control of Cuba, the Philippines, Puerto Rico, and Guam).

No further investigations took place until 1911, when the U.S. Army Corps of Engineers constructed a huge cofferdam around what was left of the *Maine*, which by then had settled considerably into the harbor floor. They removed the tons of mud that surrounded the wreck and took hundreds of photographs. After careful photo analysis, the corps also blamed the explosion on a mine. But other theories were advanced, one being that the soft coal stored in the *Maine*'s massive bunkers caught fire by spontaneous combustion (a very real shipboard hazard) and heated the ammunition to the explosion point. Proponents of this theory claimed the inward bending of the hull plates could have been caused by a tremendous inrush of water returning after the blast. Sometime after this investigation, the wreckage was cut up, and the pieces were disposed of at sea.

In 1976, U.S. Admiral Hyman Rickover, "the father of the nuclear navy," commissioned a new study of the *Maine* based on the photographs from 1911. His naval engineers concluded that the explosion actually originated *inside* the doomed battleship, but they could not prove a coal fire was to blame.

Then, in 1997, the National Geographic Society commissioned yet another review of the *Maine*'s sinking, to be published in time for the tragedy's centennial. Using the latest techniques in computer modeling, experts reconstructed the sequence of events for two highly plausible scenarios: one based on the detonation of an external mine, the other on an internal explosion caused by a fire in the coal bunkers. Despite this analysis using new technology, no definite cause of the explosion could be pinpointed.

The USS *Maine*'s sinking remains an unsolved case.

* * *

Though the circumstances surrounding the sinking of the USS *Maine* remain a mystery, the painstaking work undertaken to solve the case was unprecedented. The numerous investigations, with skilled divers, specialized equipment, and advanced techniques, gave rise to a whole new kind of sleuth: the deep-sea detective.

1 Sleuthing the Depths: The Science of Underwater Forensics

Curious and enterprising individuals have been going to the ocean floor for centuries to explore sunken vessels and salvage the treasure, weapons, and other valuables hidden within them. These adventurers gained an edge in the seventeenth century when the invention of the first primitive diving bell enabled them to reach greater depths. Despite this capability, there is no record that anyone ventured below the surface to investigate the cause of a sinking until the wreck of the USS *Maine* was first explored in 1898.

Part of the challenge facing would-be deep-sea detectives was depth; even with new technology that allowed them to dive deeper, it usually wasn't deep enough. During the early twentieth century, inventors were constantly improving diving suits to increase the depths divers could safely reach. Though the invention of SCUBA, or the self-contained underwater breathing apparatus, freed divers from a dependence on air hoses and lifelines, 200 feet or so remained the practicable depth. As most of the ocean floor lies thousands of feet below that depth,[1] deep-lying wrecks continued to be unreachable, and the causes of sinkings—including evidence of potential crimes—remained undetected.

[1] The average depth of the ocean floor is estimated at 16,000 feet. This calculation takes the ocean floor's uneven surface into account; it is studded with lofty peaks and ridges and seamed with vast trenches measuring 3 to 7 miles in depth.

The manned submersible, pioneered by the redoubtable Jacques-Yves Cousteau in the 1950s, greatly extended the range of non-military underwater operations. Cousteau's *Soucoupe Plongeante*, or *Diving Saucer*, could descend to 1,350 feet with a pilot and a scientist-observer on board. The success of the *Diving Saucer* led to a proliferation of other submersibles, especially in the United States. One of the most notable submersibles is *Alvin*, owned by the U.S. Navy and operated by the Woods Hole Oceanographic Institution (WHOI). Extensively repaired and rebuilt every three years, *Alvin* can now safely dive to 14,750 feet thanks to a new super-strong titanium hull.

To aid in underwater exploration, submersibles typically carry powerful searchlights that illuminate the pitch-black depths, cameras (both still and video) that log a pictorial record, and a mechanical arm that can be equipped for a range of uses—collecting biological specimens, taking core samples of bottom sediments or rocks, and retrieving delicate archaeological artifacts from ancient shipwrecks.

Though marvelous, submersibles have limitations. They are very expensive to build ($15 to 20 million for a high-performance analog of *Alvin*) and to operate (currently anywhere up to $45,000 per day). And the rental includes not only the actual working time but also every moment that the submersible is at sea—whether on a dive, or lashed to the deck of a mother ship or a support vehicle on its way to the work site.

Another drawback of the submersible is its crew space. To maximize strength, the pressure hulls of submersibles (the areas where the humans are contained) are usually built as spheres. To economize on construction costs and prevent the vessels from becoming unmanageably heavy, these spheres are kept rather small. As a result, after an hour or so, the occupants become uncomfortable in the cramped space. They also become very chilly, as power from the batteries is needed to drive the submersible and operate its mechanical equipment and cannot be spared for a heater. The pressure hull also becomes dripping wet inside as the moisture in the occupants' breath condenses on its walls.

Finally, submersibles are designed for slow speeds. While a snail-like pace is absolutely necessary for careful scientific observation, it also means that submersibles take a long time to travel to the sea floor and equally long to return. A typical dive takes six to eight hours, most of which is spent traveling to the bottom and back to the surface. Moreover, because they are powered by batteries, submersibles have limited ranges and operating times. They cannot travel to a work site under their own power but must be transported on the deck of a full-sized ship. (The fee for renting the ship is added to the submersible's operating costs.) And if something goes wrong, the lives of the pilot and any observers may be lost. On *Alvin*'s 308th launch, for instance, the support cables failed, and the pilot and two scientists on board barely managed to clamber out the escape hatch before the little sub sank to the bottom 5,000 feet below.

Alvin first came to the public's attention early in 1966, when a U.S. Air Force tanker plane and a B-52 bomber carrying nuclear weapons collided in midair over the Mediterranean off the port of Palomares, Spain. An H-bomb fell into the ocean; recovering it was a matter of the greatest urgency. Experts feared that in time the seawater would corrode the bomb's casing and trigger a nuclear reaction, sending a deadly rain of radioactive water and sediment over southern Spain.

The approximate position of the bomb was known since a shocked Spanish fisherman had seen it fall near his boat—he thought it was half a man's body dangling from a parachute. But the bottom was a crazy jumble of steep-sided ravines, abrupt rises, and near-vertical precipices. Regular sonar could not detect an object as small as the bomb, and towed sonar could not cope with the rugged topography of the sea floor.

Faced with this problem, the U.S. Navy decided to use a team of four submersibles, including *Alvin*, to find the bomb. It took two months of searching before *Alvin* had success; it discovered the missing H-bomb, its parachute still attached, on a steep, muddy slope nearly 2,500 feet down. *Alvin*'s pilot attempted to anchor a lift line in the mud so that a larger and more powerful companion submersible, *Aluminaut*, could attach it to the bomb, but the cable pulled loose. Then *Alvin* tried to snag the bomb's parachute harness with its mechanical claw. This maneuver also failed. After several more jury-rigged attempts at retrieval, the bomb slid 300 feet farther down the slope. At last, the navy sent down an unmanned, remote-controlled underwater recovery vehicle, or *CURV*. *CURV* was essentially an open frame with floodlights and television cameras to guide its operators on the surface and a set of powerful, motor-driven jaws to grab the bomb. *CURV* managed to raise the bomb to 200 feet below the surface. There a team of SCUBA divers working near their depth limit was able to attach additional lifting cables for the last stage of the recovery.

CURV was an early example of the ROV, or remotely operated vehicle. Popularly called a *robot sub*, an ROV is actually neither. It is not a true submarine because it has no enclosed hull. It is simply a framework that carries motors, a control unit, and other electronic equipment, plus whatever cameras and manipulative paraphernalia the mission directors choose to load for a particular mission. And it is most definitely not a robot because it cannot operate on its own. It is controlled by humans on board a mother ship on the surface, and it gets its signals and its electric power through a slender cable called an *umbilical* or a *tether*. Typically, one operator pilots the ROV while the other controls the cameras, mechanical claws, and other equipment.

ROVs have a number of advantages over manned submersibles. To begin, they are significantly cheaper to build and operate. Since they do not carry humans into the depths, they need no pressure hull. Beside the vehicle itself, a typical ROV includes a tether, a winch, a handling system, control and maintenance vans, spare parts, a generator, and special tools. The price for the complete package ranges from $10,000 for a light, nonindustrial ROV to upwards of $5 million for a deep-water salvage-capable model—a good deal less than the $15 to 20 million it would cost to build a new *Alvin*. Currently, the cost of renting an ROV can run from $3,000 to $20,000 per day, depending on factors such as its size, the complexity of the accompanying equipment, and the number of operators it requires.

Another advantage of ROVs is their ability to undertake risky missions—exploring the inside of a wreck's hull, for example. It is too dangerous to perform such maneuvers with humans on board since it is all too easy for any submersible or ROV to become wedged fast or entangled. Unlike manned submersibles, whose power and air supplies are limited, ROVs can stay submerged and on the job around the clock. And the operators sit in relative comfort in the van on the mother ship's deck rather than in the cramped and near-freezing pressure hull sphere. The renowned Dr. Robert Ballard, finder of the *Titanic* and veteran of a multitude of dives, says he would much rather be dry, warm, and with room to stretch out on the deck of the support vessel than be chilled and cramped in the pressure hull of a submersible.

ROVs are also incredibly versatile. They are used every day to inspect offshore oil-drilling rigs and undersea pipelines. With the proper manipulators, they can insert bolts and tighten them down (or the reverse); attach shackles and lifting cables to wreckage so that it can be winched up to the surface; cut metal; weld steel; drill bedrock for core samples; dig sea-floor trenches for pipelines and communication cables; and gently lift archaeological objects from the sediments in which they have been buried for centuries.

Many materials go into the construction of an ROV. The framework is either aluminum or steel; to guard against corrosion, the aluminum is coated with sealer and paint, while the steel is protected by zinc anodes. (Sometimes PVC tubing is used for light, non-industrial ROVs.) Manipulators and other special instruments may be crafted from titanium or other very durable materials, and the electronic components, sealed in waterproof casings, are made from silicon and ceramics with powdered-metal leads. Syntactic foam provides near-positive buoyancy, so an ROV that weighs more than a ton on land weighs just 100 to 200 pounds in the water. The slight negative buoyancy keeps the ROV from having to fight its way to operating depth; the light submerged weight reduces the load on the tether.

An ROV's size depends on the type of work it is designed to do. Deep Ocean Engineering's *Firefly*, which is designed to inspect the underwater portions of nuclear power plants, is about the size of a microwave oven and weighs only 10 pounds. The *VideoRay*, frequently used to locate the drowned victims of boating accidents and cars and trucks in deep water, measures 14 × 9 × 8.5 inches and weighs a mere 8 pounds. The navy's *Deep Drone 7200*, used for salvage-and-recovery work to a depth of 7,200 feet, measures 9.25 × 4.6 × 6.17 feet, roughly the size of a subcompact car turned on its side. (The navy does not give its weight). *CURV III*, another navy salvage-and-recovery vehicle, has a depth capability of 20,000 feet and weighs approximately 6.5 tons. It can lift 2,500 pounds. The navy does not furnish its dimensions to the public.

ROVs were originally developed by the U.S. Navy in the 1960s to recover spent practice torpedoes and dummy bombs from training areas off the California coast. The oil-drilling industry soon recognized their value for a variety of underwater tasks and developed them further. To this day, ROVs are mainly used in offshore oilfields. However, they have also been found useful for locating and recovering sunken ships and crashed aircraft, as documented in later chapters.

Although ROVs are the workhorses of undersea investigation, there are times when a diver still does the job better. For delicate, close-up maneuvering, the human hand still cannot be beaten. When visibility is bad, it may be necessary to photograph evidence at very close range. Here again the diver is superior to the remotely controlled vehicle.

Special breathing mixtures enable divers, whether helmeted or SCUBA, to go beyond the usual 200-foot working limit for brief tours of duty. But for deeper water, the armored diving suits first developed in the late 1920s and early 1930s have come into their own. The great advantage of these armored suits is that the air inside remains at normal surface pressure; when surfacing, the diver does not have to undergo a lengthy decompression process. The technical term for these suits is *monobaric*, or *atmospheric diving suits* (ADS, for short). The ADS has worked well at depths to 1,000 feet, and greater depth capability is on the horizon. Looking much like swollen bugs from science-fiction illustrations with domed visors and backpacks attached, these rigid-bodied suits permit the movement of the diver's arms and legs by means of rotating joints. This advance was made possible when high-tech materials were developed to seal those joints from the crushing pressure of the water. There are different variants on the armored suit: the JIM suit, the Newt suit, the SAM suit, and the WASP suit. The WASP is even more surrealistic than its competitors. Without jointed legs—or any kind of legs at all—it resembles a man-sized pupa with arms. It has been upgraded to a depth capability of 2,000 feet.

For working with tools, all these rigs have mechanical grabbers where their "hands" should be; the diver operates this machinery from inside the suit. While the other rigid suits permit the diver to walk on the bottom, the WASP suit is meant to "fly" in the water, maneuvered by a set of electrically powered thrusters. The JIM suit has been used to retrieve valuable anchor chains lost by tankers at 375 feet and salvage a crashed helicopter off Plymouth, England, at a depth of 325 feet. The WASP suit even appeared on TV's *60 Minutes II* after a diver used it to recover a sunken hoard of Nazi counterfeit money from the 348-foot-deep Lake Toplitz in the Austrian Alps. An ROV located the money—fake British ten-pound notes—at 200 feet. After soaking for 55 years, the paper was so disintegrated that the ROV's mechanical claws could not grasp it. But a diver in a WASP was able to recover a sample large enough for conclusive forensic analysis.

Another technique used is saturation diving, based on the principle that once working depth is achieved the amount of gases dissolved in a diver's blood does not increase, regardless of the length of the dive. In practice, the diver might be exhausted and chilled to the bone after as little as fifteen minutes of work. But with saturation diving, the diver returns to a pressurized habitat on the sea floor; there it is possible to strip off heavy, cumbersome gear, warm up, take a shower, read, listen to music, eat, and sleep before going out to work again. Experience has shown that a saturation diver can work four times longer than a conventional diver at a given depth and accomplish much more. Another advantage of saturation diving is that the diver does not have to undergo the stress of decompression after every sortie. American businessman-investigator Gregg Bemis has used saturation divers in his explorations of the World War I British liner *Lusitania*.

At a depth of 300 feet, a diver can use the habitat for up to two weeks before returning to the surface to go through decompression in a shipboard chamber. Two weeks is the limit because long exposure to high pressure does nasty things to the human body: bones are eventually weakened, and the nervous and circulatory systems can be damaged, for example. A variation on this system is to house the habitat on the support ship's deck; the diver makes trips to and from the bottom in a small pressure chamber and enters and exits the habitat through an airlock much like the ones in space labs.

How great is the pressure of the water, exactly? At sea level, the pressure of the atmosphere is approximately 14.7 pounds per square inch, the weight of a column of air reaching from ground level to the topmost fringes of the atmosphere. But water is much denser than air, and at a depth of only 33.8 feet the pressure is two atmospheres. Another atmosphere's worth of pressure is added for each 33.8 feet descended. At 1,000 feet below, the pressure is equivalent to 435 pounds pressing on every square inch of a body or an object, as if it were being squeezed by a giant's hand. At 13,460 feet, the depth at which the British bulk carrier *Derbyshire* lies, the pressure is 2.9 *tons* per square inch. Such pressure is difficult even to imagine.

Pressure is the great enemy of all undersea work, posing dangers to submersibles, ROVs, and divers alike. The pressure of the depths accelerates the speed at which metal is corroded by salt water, magnifying hidden flaws and sometimes causing cracks in the vessel and its equipment.

For divers, the dangers are multiple. They must be supplied with air whose pressure is equal to that of the surrounding water at any given depth. Otherwise, the pressure of the water prevents divers from breathing even at a relatively shallow depth and crushes him or her at greater depths. There have been numerous cases of "squeeze," in which the diver's suit suddenly loses pressure, causing the luckless diver's body to be forced up into the suit's rigid helmet by the external pressure of the water. Needless to say, no diver has survived this experience.

In another scenario, when the diver breathes ordinary air under pressure, nitrogen dissolves in the blood and body fat. Should the pressure be lowered suddenly (if the diver returns to the surface too quickly, for example), the dissolved nitrogen comes fizzing out like the gas in a bottle of soda that has been shaken. In this condition, called *the bends,* nitrogen bubbles block blood vessels and lodge in the joints, causing agonizing pain and contortions and sometimes unconsciousness and death. And nitrogen presents yet another peril to the diver: *nitrogen narcosis,* or rapture of the depths. Here the diver becomes woozy and disoriented and loses touch with reality. SCUBA divers suffering from nitrogen narcosis have been known to swim down into the depths when they meant to return to the surface.

Jacques-Yves Cousteau attempted to avoid this danger by breathing pure oxygen. But he found that, under pressure, oxygen is a violent poison. So deep-sea divers still needed a replacement for nitrogen. Helium was one answer. This gas dissolves very little in the body, and a helium-oxygen mixture, or *heliox,* came into vogue. This combination works safely but with two drawbacks: the diver gets chilled more rapidly, and the voice is affected so that it sounds like Donald Duck's. At 1,000 to 1,500 feet, helium becomes dense enough to adversely affect the diver's breathing; some divers compensate for this problem by using a hydrogen-oxygen mixture, or *hydrox,* despite its dangerously explosive nature. A more recent development is *trimix,* a blend of oxygen, helium, and nitrogen that the diver can adjust to suit changes in depth. The deeper the body goes, the less oxygen it requires, so divers will dilute the helium-oxygen mixture with nitrogen as required, despite the danger of the bends. Since helium can also form bubbles in the bloodstream, extreme care must be taken when diving to greater depths.

With all of the tools and techniques described above at their disposal, deep-sea detectives are called upon when a tragedy at sea (or in a deep lake) must be investigated. Government safety agencies want to establish the cause of a shipwreck in order to prevent future accidents, and insurance companies want to make sure that foul play and fraud were not involved. Since governments can afford only limited investigative corps, and insurance companies cannot afford them at all, such operations are typically contracted out.

In a typical search-and-recovery operation, the wreckage is initially located with sonar, a piece of equipment without which locating deep-lying objects would be impossible. Fittingly, the invention of sonar was inspired by a marine tragedy: the 1912 sinking of the *Titanic.* The Canadian-born Reginald Fessenden, who had worked for Thomas Edison in the United States, realized that the amount of time it took for a burst of sound to bounce back off an object gave a measure of distance. (It was already known that sound travels at a fairly constant speed.) This echo principle even worked at night or in heavy fog, when visibility was nil. Fessenden's original "pinger" was nothing more than an oversized buzzer, but if the *Titanic* had been equipped with such a device the officers and crew on duty could have avoided colliding with the fatal iceberg.

Water is an even better conductor of sound than air, and by 1914 true sonar was invented. A pinger mounted underwater on a ship's hull sends out periodic bursts of sound, which are reflected off solid surfaces and picked up by hydrophones, or waterproof microphones, also on the hull. Another device converts the elapsed time into the ship's distance from the target. During World War I, England and France used sonar to locate lurking German submarines. By the mid-1920s, most of the world's navies were using it. Oceanographers began using sonar, too, since it turned out to be an excellent depth finder—much more accurate than the traditional method of using a weighted line. Its precision made it useful for mapping the ocean floor. Though sonar can be "fooled" by a number of things, such as fluctuating water temperatures, turbulence, turbidity, changes in salinity, and even layers of plankton or large whales, it is still the most reliable tool available for its purpose.

In time, sonar came into use for locating wrecks. Sonar can look straight down, forward, or to the sides. At present, side-scan sonar is the first tool employed by a search-and-recovery team. In this type of sonar, the sound-wave emitters face outward at an angle so that they cover a wide swath of sea floor, from about 1,640 to 4,920 feet on either side of the carrier. Sonar follows a rule of thumb: the lower the frequency, the larger the range, and the less distinct the resulting image. Low-frequency sonar (also called *low-resolution sonar*) is useful for locating large objects, but for fine detail high-frequency (or *high-resolution*) sonar is required; it covers a much smaller area—about 330 to 3,300 feet. The low-frequency sonar is often used for the initial search; then high-frequency sonar is used for positive identification. Some claim high-frequency sonar is able to read the raised lettering on an automobile's license plate from a distance of 200 yards.

The sonar may be mounted either on the support ship itself or on a *towfish* (also called a *sled*), which is actually a sonar-carrying framework towed behind the ship. The sonar images are relayed to the support ship's control center, where they are recorded on paper and also converted to visual images by sophisticated computers. Once a likely object is found on the bottom, an ROV armed with side-scan and forward-scan sonar plus TV and still cameras is sent down to investigate. The ROV homes in on the presumed wreck and sends data back to the control center, which is located in a van on the support ship's deck.

If the object is indeed a wreck—and the right one[2]—the ROV begins the process of collecting evidence, which usually takes more than one dive. To facilitate an easy return to the wreck, the ROV will drop a pinger on the bottom near its target to signal the precise position.

The ROV's first task is to take hundreds of still photographs[3] and shoot extensive film footage of the wreckage and its distribution. These images are then evaluated in a shoreside laboratory. Naval architects and engineers pore over images of the wreck itself and from the distortion of the metal they calculate how the ship or aircraft failed. Images of the wreckage's distribution—which can number in the hundreds or sometimes thousands—are painstakingly pieced together into a mosaic of overlapping shots. An analysis of the resulting composite image can tell the photo interpreters whether the vessel, typically a cargo carrier, broke up at the surface or on its way down to the bottom. To the layperson, a steel ship is a solid, sturdy structure that looks invulnerable, but in fact its own weight, plus the weight of the cargo, impose unimaginable stresses on the hull as it sinks, and few wrecks reach the bottom intact. Cargo may be scattered far and wide; the pattern of its distribution provides clues to the cause of the sinking.

---

[2] Not all objects on the sea floor are wrecks; the sonar may well pick up boulders or ledges. Furthermore, some areas of the sea floor, such as the White Sea off the northern coast of Russia where the submarine *Kursk* went down, are so littered with shipwrecks that the right one can be difficult to pick out.

[3] For still photographs, black-and-white film is the standard choice because it is faster than color film, making it ideal for dark conditions on the sea floor.

The investigative mission may call for retrieval of physical, as distinct from photographic, evidence. For example, a piece of hull plating with the ragged edge of a fracture may be brought to the surface for inspection. In this case, the ROV cuts the piece loose with a torch or abrasives, then attaches the tortured metal to a toggle on the end of a cable. The evidence is then hoisted to the surface by a winch or crane on the support vessel. Alternatively, the item recovered may be a section of piping, scorched wiring, or seat-cover fabric from a downed airliner. Small items such as these may be placed in a mesh basket, which is then winched to the surface. ROVs themselves do not have a great lifting capacity, although in some cases they will bring evidence to the surface in a basket they carry on board.

The recovered physical evidence—say a piece of distorted hull or hatch cover or the twisted hinge of a loading ramp—is first examined for what the experts call *gross morphologies:* that is, changes in structure visible to the naked eye. Is the metal bent inward or outward? Is it torn? dented? holed? rusted through? The next step is to put the sample under the microscope. But first it must be cleaned. A clean surface is typically achieved through *electropolishing,* a process best described as electroplating in reverse. Where electroplating adds a thin layer of metal to the object being plated, electropolishing removes a thin layer of metal from the object being polished, and the process takes off surface impurities and dirt and leaves a smooth surface—smoother than if the object had been polished by machine. The sample is then treated with nitric or sulfuric acid to bring out the grain structure.

Next an expert looks at the object under an electron microscope (an optical microscope does not give high enough magnification) and determines whether the failure was ductile or brittle. In a ductile failure, the metal is literally pulled apart—it looks irregular under the microscope. When a part is exposed to tension greater than it is designed to handle, a ductile failure results. With a brittle failure, the metal looks flat under the microscope. This type of failure occurs when accumulated stresses weaken the metal until it breaks.

For those who know what to look for, the size of the metal grains can also tell a story. Basically, the smaller the grains, the stronger the metal. Under repeated stress, the small grains gradually coalesce into larger grains, which weaken the metal. "Large," of course, is a relative concept, and even large grains cannot be seen with the naked eye. In the laboratory, metal samples may also undergo mechanical tests to determine their strength and resistance to deformation and chemical analysis to determine whether they contain the proper elements in the proper proportions. The type and extent of corrosion are also evaluated.

If already corroded, metal samples must be protected from further deterioration while making the journey from the wreck site to the lab. It is essential to prevent oxygen from coming into contact with them. To this end, a sample is typically placed in a container of deaerated water (deaeration removes oxygen and other dissolved gases from water, rendering it much less chemically active). Sometimes the sample is packed in a vacuum container instead. The depth at which the wreck lies affects the rate and extent of corrosion. Deep water contains less oxygen than surface water and is therefore less corrosive. However, iron-eating bacteria can invade a wreck even in deep water, as they have with the *Titanic.* These bacteria produce *rusticles*—icicle-like drip formations of mushy iron compounds—which slowly destroy the steel.

Search-and-recovery operations, in addition to collecting and analyzing evidence from the depths, involve slow and deliberate work. The process can take weeks, months, or longer—but it usually yields conclusive results, not to mention healthy doses of danger and drama.

One of the most fantastic stories in the history of deep-water search-and-salvage operations is Project Jennifer, which involved the recovery (on behalf of the CIA) of a Soviet Golf-class submarine that had sunk in some 16,500 feet of water 750 miles northwest of the Hawaiian Islands. Soviet subs of this class were diesel-powered, but this particular one carried the latest nuclear-tipped missiles. It was of the greatest importance to U.S. intelligence to secure the wrecked sub and harvest its secrets; it was equally important to the Soviet Union to prevent the sub from falling into the hands of the "capitalist hegemonic power."

Though they searched frantically for it, the Soviets had no idea where their lost submarine was. The United States had a fairly close approximation of its resting place, however; its listening stations in the Pacific had been tracking the vessel and noted the area in which its transmissions ceased. The U.S. spy sub *Halibut*, supposedly used for oceanographic research, was deployed to the mid-Pacific location deduced by U.S. intelligence. *Halibut*'s sonar, in combination with a sonar sled towed by the navy surface vessel USS *Mizar*, determined the sub's precise location.

Although the CIA was anxious to get hold of the codebooks, missiles, and other paraphernalia K-129 (the identification number of the lost Soviet sub) was carrying, it waited six years before acting, confident that its Soviet rivals would not get there first. In the interim, the navy contacted the reclusive billionaire Howard Hughes and convinced him to commission a specially built search-and-recovery ship. The vessel, quite unlike anything that had ever been seen before, was named the *Glomar Explorer*. It measured 619 feet in length and 116 feet in breadth at its widest point. Its horsepower and speed remain classified. In the middle of the hull was a huge "moon pool" surmounted by a tall derrick similar to the ones used in drilling for oil. The derrick would lower a giant five-fingered grab to clutch the submarine and then raise it to the surface.

The *Glomar Explorer* was to operate under a cover story that it was mining manganese nodules from the depths of the Pacific Ocean. There was, in fact, a good deal of interest in these sea-floor mineral deposits at the time. Its real purpose, of course, was quite different. The ship was built under unbelievably strict secrecy. According to one account, any government employee who visited the construction area had to assume a false name and change into a disguise at a safe house some distance from the project. Additionally, one of the planning offices could be reached only through a secret door concealed by a wall locker on wheels.

In 1974, President Gerald Ford authorized the mission, and the *Glomar Explorer* proceeded to its station without incident. To lend authenticity to the cover story, manganese nodules from the Blake Plateau off the southeastern coast of the United States were handed out to the crew's families and friends. The strange-looking ship arrived at its station and began its recovery operation on July 4,

1974. Guided by sonar and kept in position by thrusters (the ocean was far too deep to permit anchoring), the *Glomar Explorer* slowly lowered a long string of oil-field drill pipe through its derrick. When the pipe was out of sight beneath the water's surface, CIA operatives attached the grab, whose existence had been kept secret.

The grab was close to its target when the operator lost control, causing it to strike the sea floor next to the sub's hull. Since it was so close, the operator raised the grab slightly, maneuvered it over the sunken submarine, and lowered it successfully. Very slowly, the operator began to raise the grab and its 3,000-ton payload. When the sub came clear of the sea floor, the added weight sank the *Glomar Explorer* 7 feet lower in the water.

Suddenly, a Soviet "trawler," almost certainly fitted with spying gear, appeared and started circling the *Glomar Explorer*. While the onboard mission chief pondered what action to take, the Soviets decided the Americans were probably wasting their money on another pretentious and ill-conceived project. After coming so close to the lost submarine, the Soviet trawler simply sailed off.

When the submarine with its coveted load of knowledge was just 5,000 feet from the surface, three of the grab's five claws, probably weakened by their impact with the sea floor, broke off. Only partly supported, the sub broke in two. According to the story that the CIA has chosen to share with the public, only the front 38 feet of the vessel were saved; the remainder sank back to its ocean grave. But the recovered portion supposedly held K-129's codebooks and two of its nuclear missiles, plus the bodies of six to eight crew. According to some reports, a third nuclear missile slipped from the craft and fell back to the seabed. It was a purported tense moment on the *Glomar Explorer* as the crew waited for a deadly explosion. But the warhead settled harmlessly on the ocean's bottom.

The recovered Soviet crew were given a formal burial at sea in English and Russian; the mission organizers had provided copies of the ceremony in Russian. The ceremony was videotaped and presented years later to Russian President Boris Yeltsin as a goodwill gesture. As for questions about the sub itself—how did it sink? how did the CIA preserve the salvaged intelligence material? how was that material used?—that information is known only to a privileged group of investigators. It has never been made public.

* * *

Thanks to submersibles, ROVs, and skilled divers, virtually no place in the sea is now beyond reach. For the first time in history, determining the causes of devastating maritime disasters, whether accidental or acts of sabotage, is within the bounds of underwater forensic science. Deep-sea detective work has come into its own.

2 No Match for the Forces of Nature: The *Derbyshire*, the *Gaul*, and the *Flare*

"Vessel hove to ... violent storm ... Force 11 winds ..." came the crackling radio message from the *M.V. Derbyshire* on September 9, 1980, at 11:00 P.M. These were the last words heard from the giant bulk carrier before it vanished southeast of Okinawa, Japan, without a *Mayday* call. Blindsided by an unexpected change in a storm track, the *Derbyshire* had blundered into the most violent sector of Typhoon Orchid.

Built in 1976 by the reputable British firm Swan Hunter, owned by the well-regarded British shipping company Bibby Brothers, and manned by an experienced, all-British crew, the *Derbyshire,* like her five sister ships, was a showpiece of Britain's merchant fleet. Nine hundred and sixty-five feet long (longer than three football fields placed end to end) and 145 feet across at her broadest point (as wide as a six-lane highway), the *Derbyshire* was designed to carry up to 66,363.14 long tons of oil, ore concentrate, or bulk freight. A double skin of heavy steel plates armored her sides and bottom. The rating agency Lloyd's Register of Shipping, which had studied her construction plans and inspected the finished work, pronounced her fit for service on the high seas. Indeed, the *Derbyshire* appeared large and sturdy enough to withstand anything short of a torpedo or a mine.

The *Derbyshire*'s single screw, driven by a diesel engine developing 30,400 horsepower, propelled the gigantic ship at a cruising speed of 10 knots—far from swift, but there are limits to the speed at which a load of 173,192 tons of cargo, plus the ship's weight of 91,654 tons—close to 3 million tons in all—can be propelled. And the kinds of cargo that she carried were not perishable, so speed was not a prime factor.

On her last voyage, the *Derbyshire* was carrying a cargo of iron-ore pellets from Quebec, Canada, to Kawasaki, Japan. Four days passed in silence after the ship's last transmitted message, but the *Derbyshire* did not appear at her destination. The worried owners contacted the Japanese Maritime Safety Agency (MSA). But under Japanese regulations, the MSA was not permitted to undertake a search for a lost ship until twenty-four hours after its scheduled arrival date (in the *Derbyshire*'s case, September 14). On September 15, the MSA sent out two patrol boats and two aircraft to the *Derbyshire*'s last known position. There was no sign of a vessel, but about 40 miles from the site they spotted an oil slick about a mile long and half a mile wide. They assumed that the oil was seeping up from the ruptured fuel tank of a ship on the sea floor. Two follow-up searches also found this upwelling oil. Six weeks after the *Derbyshire* was lost, a Japanese tanker found an empty lifeboat bearing her name almost 700 miles west-southwest of the presumed wreck site, the last relic of the unfortunate ship. Forty-two officers and crew, along with two wives, were lost with her.

The shipping world was mystified. How could such a large vessel sink in a storm at sea? Why did the crew not send out an emergency call, especially since the ship was equipped with an automatic *Mayday* device that worked at the push of a button? The answers would be a long time coming.

The initial government inquiry, conducted by the Admiralty Court, simply concluded that the ship had succumbed to the force of the storm. There was no way that the technology of 1980 could discover a wreck nearly 13,500 feet down, much less gather evidence to determine the cause of its sinking. There the matter rested for eighteen months until a sister ship of the *Derbyshire*, the *Tyne Bridge*, developed alarming cracks in her deck plating at Frame 65 and had to return to port. (In shipbuilders' parlance, a frame is one of the transverse ribs to which the plates of the hull are fastened.)

For the *Derbyshire* and the rest of her class, Frame 65 marked the end of the cargo space and the beginning of the engine room, about four-fifths of the way from the bow to the stern. A bulkhead, or partition, separated the spaces. Unlike those of the luckless *Titanic*, the *Derbyshire*'s bulkheads stretched all the way from the bottom to the top of the hull so that each cargo hold was watertight—a precaution against massive flooding of the entire hull.

The families of the dead crew, represented by the *Derbyshire* Families Association (DFA) and the International Transport Workers' Federation (ITF), suspicious of the shipbuilder and the owners, called for an investigation of the wreck. Mariners have had good reason to doubt the seaworthiness of many merchant vessels, which are ill maintained by owners whose sole concern is maximizing profits. Though the *Derbyshire* and its owners did not seem to fall into this category, the families and the unions wanted to be sure.

But the case remained closed for another six years. Then, in 1987, another of the *Derbyshire*'s sister ships, the *Kowloon Bridge*, went on the rocks off the coast of southern Ireland; it broke up at Frame 65 after several days. This development aroused fresh suspicion of structural defects, and the British government felt compelled to undertake a formal investigation of the *Derbyshire*. Reviewing the known facts about the ship, its cargo, and Typhoon Orchid, the investigating board concluded that the probable cause of the sinking was failure to secure the front hatch cover properly, which allowed storm waves to flood the front cargo hold.

The DFA was outraged by this implication that the crew had brought their fate on themselves through negligence or incompetence and mounted a large-scale protest in the press. Since the government was not likely to reopen the investigation (the DFA and the ITF suspected that it was protecting the capitalists), in 1994, the ITF funded its own expedition to find and identify the wreck. Ideally, it wanted photographic evidence of the damage. For this job, it chose an American firm, Oceaneering Technologies, an organization with an impressive record of locating deep-water wrecks. Among Oceaneering's finds were the notorious *Lucona*, the space shuttle *Challenger*, and a number of airplanes.

For the *Derbyshire* mission, Oceaneering flew its search equipment to Yokohama, Japan, where it had hired a Japanese support ship, the *Shin Kai Maru*. Now, it takes a certain amount of time to set up a ship for such an expedition. The winches that raise and lower the towfish and ROV must be set up on the deck. A spot must be found for the control van, which needs to be securely tied down. All the electrical connections must be made and tested thoroughly. The mission encountered an additional delay when diligent Japanese customs officers took two days to clear the mountain of equipment that Oceaneering had brought into the country. Then the hydraulic power unit on one of the winches used to control the towfish overheated during a test, and a replacement part for its diesel engine had to be secured from Tokyo. The *Shin Kai Maru* was several days behind schedule before it was able to set out on May 26.

The weather turned the following day, sending the *Shin Kai Maru* into a tropical depression like the one that had spawned Typhoon Orchid. But by the time the expedition reached the search area (based on the reported position of the oil slicks), conditions had improved enough to allow for some trial runs. The plan called for sonar scans along seven lines, each measuring 14 nautical miles in length. These lines were plotted about 2 miles apart. The sonar swath would cover an area just under 3 miles, giving sufficient overlap to cover the region's entire seabed.

The towfish, a huge *Ocean Explorer 6000* fitted with side-scan sonar, was carefully lowered over the side, and the search began. Back and forth the *Shin Kai Maru* chugged at a cautious 2.5 knots (just under 3 miles per hour), dragging the ponderous towfish behind 5 miles of cable and at an altitude of 1,500 feet above the sea floor.

On the first two actual runs, the sonars picked up a strange image on the sea floor: two "targets" with squarish corners, which suggested that they were not geological features. However, the Oceaneering experts assumed that the *Derbyshire* would be in one huge piece, so they were dubious about the find. Then another sonar run picked up a big target in about the same area, an encouraging discovery.

At 1:23 A.M. on June 3 (these searches are conducted around the clock), the sonar, now on high resolution, transmitted a significant concentration of images. The team pored over these pictures until

3 A.M. and decided they represented a huge field of debris that could be consistent with a very large ship and its spilled cargo of iron-ore concentrate. The position was 25°86' N, 133°53' E.

The discovery was very encouraging, for time was running out on the *Shin Kai Maru*'s rental contract. The owners of the support vessel were also anxious to get her back to port before the typhoon season set in in earnest. The next step was to send down the ROV, *Magellan 725*, to get a closer look. But the weather worsened again, and the search team could not risk having the ROV crash into the side of the ship while it was being lowered into or raised from the turbulent water. Not until June 7 did the weather permit the crew to place *Magellan 725* in the water.

Magellan 725 was launched at 9:20 P.M. It did not reach the bottom until 1:40 the next morning. One hour and six minutes later, the ROV sent back the first visuals of the wreckage, which lay almost 13,450 feet below the surface. Soon, a number of large pieces of wreckage came into view; at 5:32 A.M., the ROV's cameras found the bow, which had broken off from the rest of the ship. On the port side, part of the ship's name, ——*SHIRE*, was visible. This was the most conclusive evidence that Oceaneering could hope to gather within the mission's time allotment, so *Magellan 725*'s manipulator arm laid a memorial plaque beside the spare anchor on the forecastle of the *Derbyshire* and took a sample of the shimmering iron-ore sediment that covered so much of the sea floor. The stern section of the luckless bulk carrier was not conclusively identified on the dive.

From the distribution pattern of the pieces of the wreckage and of the gravel-like cargo, Oceaneering and the ITF concluded that the *Derbyshire* had broken up at or near the surface and subsequently imploded, scattering hundreds of fragments over a wide area. The ship obviously sank rapidly, since the crew had no time to send a *Mayday* message. Still unanswered at this point were two questions: (1) how did the implosion occur? and (2) did the stern section actually crack at Frame 65?

"Many questions about how the *Derbyshire* came to be so completely destroyed remain to be answered," said the ITF report, which called for a new public inquiry by the government. The issue was not only a moral one, claimed the ITF, but also a practical one, since there was still no explanation for the cause of the *Derbyshire*'s foundering—and a large number of bulk carriers had foundered over the previous two decades, with great loss of life.

Succumbing to public pressure in early 1995, the British secretary of state called for a new investigation of the tragedy, to be conducted as a two-phase survey. Phase I, accomplished in July 1996 using a "ship of opportunity" (in fact, a vessel chartered by Oceaneering, which happened to be in the vicinity), located the stern of the *Derbyshire* among the wreckage in an area with excellent visibility. Luckily for the next phase of the survey, the wreckage was not deeply buried.

For the crucial Phase II, which took place in March and April 1997, the British government needed a more specialized organization. It approached the U.S. government for assistance, and that government passed on the request to the U.S. Navy. The navy contacted the prestigious Woods Hole Oceanographic Institution (WHOI), based at the southwestern corner of Cape Cod, Massachusetts. WHOI's normal business is conducting scientific investigations of such subjects as the strange life-forms found in the Pacific's deep-sea hydrothermal vents, the Sargasso Sea ecosystem, and the geology of plate tectonics, as well as measuring ocean conditions and mapping the ocean floor. But it also has the equipment and expertise to find and examine deep-lying wrecks, and it has undertaken a number of such missions for the U.S. Navy.

Britain's Department of the Environment, Transport, and the Regions (DETR) had laid out a far-reaching plan for WHOI. The objective was to conclusively prove or disprove each of fourteen possible scenarios, some of which included events the crew of the *Derbyshire* could not have foreseen—"since the sea most often springs surprises," reasoned the DETR. Disproving a scenario was just as important as proving it, for such a rigorous process was the only way to eliminate rumors, misleading conjectures, and accusations of conspiracies.

DETR, wise from experience, allotted forty-seven days for gathering photographic evidence from the wreck—an unusually long time. To get an overview of the wreck and its scattered pieces, WHOI provided the celebrated towfish *ARGO*, armed with side-scan sonar and digital imaging equipment; for close-up camera work, it supplied the ROV *Jason*. The support vessel was a dedicated oceanographic research ship, the University of Washington's *Thomas G. Thompson*.

The *Thompson*'s voyage began in Guam, where the ship took on fuel (an all-night procedure) and provisions. A WHOI team spent several days putting *ARGO*, *Jason*, and other equipment through extensive testing. It took a good deal of attention to get the ROVs into top shape: the crew tested electrical connections and responses to controls, aligned the sonar transducers, and so on. During this process, the air-conditioning in the control van went off, and the hot, steamy atmosphere began to weep onto the computer monitors, completely obscuring them. Eventually, the air-conditioning was restored to working order, the monitors were allowed to dry out, and work proceeded.

When all the instruments were working to the team's satisfaction, they were secured with bungee cords to prevent them from being tossed about and damaged on the rough seas, which assailed the ship as soon as it cleared the harbor. It was March 9 when the *Thompson* set out. On board was a young technician named Wil Howitt, who kept a vivid and entertaining journal of the expedition. About the *Thompson*'s departure Wil Howitt wrote, "It's quite something to see the 300-foot vessel, which has been as stable as a city block for the last week, being tossed around like a toy."

Using the coordinates provided by Oceaneering's global positioning satellite (GPS), the *Thompson* reached its station, the approximate site of the *Derbyshire* wreck, late at night on March 14. The crew's first task was to drop transponders to the sea floor. A transponder is a spherical device the size of a large beach ball (about 2.5 feet in diameter) that receives sonar signals and sends answering pings back to the surface. Taking factors such as water currents, temperature, and salinity into account, a computer converts the amount of time elapsed between the sent and received signals into a measure of depth. The ship spent most of the day "steaming" around the transponders, testing their responses with its hull-mounted sonar and getting a fix on their positions.

That evening the crew carefully lowered a sonar-equipped sled into the water. Dangling behind some 2 miles of cable (there were

5.5 miles of cable in all, which weighed about 16 tons), the towfish took its first run the next day. Oceanographic cable contains steel for strength (in this case, 15,000 pounds of pull); inside the steel sheathing lie copper wire and fiber optics for telemetry. That this cable is only 1 to 1.5 inches thick is a small marvel of technology in its own right. While the towfish is on a dive, the winch is manned by a skilled operator who reels the cable in and out to keep the towfish at the right height from the sea floor and away from obstacles that show up on the sonar screen.

The *Thompson* traveled back and forth in a pattern that oceanographers and wreck seekers call *mowing the lawn.* Each pass covered about a half mile of sea floor and took about an hour. This slow speed is necessary so features on the bottom are not missed. Because of the length of the cable, it took two hours at the end of each pass for the ship to turn around. Why not haul in the towfish and drop it back in the water instead? Apart from risking costly damage to the sled should it hit the hull or railing, a cable can snap under the stress—and its flailing end is capable of cutting a bystander in half. So safety calls for making a long, slow turn despite the expense of extra time and fuel.

Working in eight-hour shifts, three court-appointed British assessors sat in the control van with the towfish and ROV pilots and the scientists who interpreted the data. Dana Yoerger, the Woods Hole scientist who had designed *Jason,* was present as a troubleshooter. His WHOI colleague Jonathan Howland was in charge of constructing photo mosaics from thousands of digitized still photos.

On March 16, *ARGO* replaced the first sled, and digitized photo images started coming in over the cable. The sharpness of these images delighted the British assessors and the rest of the team. In addition to the digital cameras, *ARGO* "saw" through two video cameras (one that looked forward and one that looked down at an angle) plus front-scan and side-scan sonar units. Even at this early stage, the leading British assessor, Robin Williams, was able to identify some of the twisted pieces of the wreckage.

A couple of days later, *ARGO*'s cameras picked up deep gouges in the sea floor, almost certainly created by Oceaneering's search equipment. This find led to lots of jokes about "side-slam" sonar instead of "side-scan" sonar. Apparently, Oceaneering's winch operator had not been vigilant about keeping his expensive sonar sled at the proper height above the ocean's bottom.

On March 19, as Wil Howitt operated the winch that controlled *ARGO*'s altitude above the sea floor, he got his first glimpse of the *Derbyshire*'s stern. The navigation team had told him to expect it, so he was prepared when it loomed up on *ARGO*'s front-scan sonar, dead ahead. It looked to be about 66 feet tall, so he was forced to pull the costly and irreplaceable camera sled up rapidly to clear its height. The stern of the wreck, he noted in his journal, was bristling with obstacles on which the tether could be snagged, including a huge crane that had once helped the *Derbyshire*'s crew to load and unload cargo.

On March 20, a storm began brewing. If the waves got too high, *ARGO* would have to be pulled up and secured on deck, a process that normally took a good three hours. The leadership decided to take a chance, however, and continued towing *ARGO* over the wreckage field. Luck was with them, and the storm subsided.

Every research voyage is prone to problems, and on March 23, *ARGO*'s telemetry system, which ran the flow of data from the sled to the ship and vice versa, quit cold. The searchers were left with no way to tell how high *ARGO* was above the sea floor, in what direction it was heading, or what it was seeing. The blinded camera sled was reeled in and repaired. During its downtime, the WHOI team worked on developing a system to break the wreckage field into tiles, or squares, measuring 328 feet on each side. Images from the wreckage field, which measured just under a mile on each side, could then be sorted by tile. If the assessors wanted to look at a particular piece of wreckage, it would be relatively easy to locate using this grid system.

On March 26, *ARGO* suffered another breakdown: several fuses in the shipboard power unit blew, and *ARGO*'s forward thruster went dead. More time was lost as the sled was hauled up and fixed. And the expedition's woes weren't over. Two days later, the GPS system failed, apparently the result of a recurring software problem. This glitch wasn't an easy one to iron out. Since the civilian GPS resolution of 164 feet is too coarse for delicate work like mapping a wreckage field, the *Thompson* was using a special military version that required an access code—and the code's authorization had apparently expired. Outside of U.S. waters, military procedures require that the replacement code be carried, *by hand,* by a person with special security clearance. It looked as if the team would have to arrange for the code to be carried to Okinawa by an admiral; the *Thompson* would steam in and pick it up there. The process would mean lost charter time that could not be made up.

In the meantime, the team struggled on with a makeshift grid system that combined the accurate but slow position readings from the transponders on the sea floor with the coarse but rapid readings from the civilian GPS. The autopilot did not work with this system, so the ship had to be steered manually. Some of the people who drew duty at the helm could not steer very straight. Wil Howitt noted that, by looking at the red lines on the control-van monitor, which marked the ship's progress, he could tell who was handling the controls.

Next, *ARGO*'s stern thruster failed, causing general consternation. *ARGO*'s "eyes" showed what appeared to be a rope floating upward from some sort of anchor on the bottom. This rope had apparently fouled the thruster, and it could not be untangled without bringing *ARGO* back on deck. A quick decision had to be made: if the rope were too securely attached, *ARGO*'s tether might snap under the load; if it were debris from a lost fishing net, it might have a hollow glass float on its upper end. This float could be dragged down until it hit ARGO and imploded, causing further damage.

Howitt, at the winch, hauled *ARGO* up about 330 feet and managed to reposition the sled so that its downward-looking camera could survey the scene. The camera showed several strands of a severed rope twisting in the current. When *ARGO* was safely on deck, the team found about 330 feet of a slender, orange-for-visibility line wound around the thruster prop, which was nearly yanked out of its socket. The rope was later identified as the *Derbyshire*'s, a message line that would have been sent by rocket to another ship to haul over a towline.

Restored to working order, *ARGO* went back in the water shortly after midnight on March 31. On duty again, Wil Howitt, along with a couple of shipmates, caught some squid that were attracted by the ship's operating lights. Unfortunately, they were infested with parasites—the crew went without a midnight snack. *ARGO* spent much of the day exploring an anomaly on the sea floor revealed by sonar; the search team thought there was a chance that it might be an outlying piece of the *Derbyshire*. It turned out to be an old submarine lava flow.

Back at the wreckage field, *ARGO* returned to running lines to fill in any gaps between the lines already traversed. Although little additional wreckage was found, the operation was necessary to ensure that the survey missed nothing of importance.

On April 7, a Japanese shuttle vessel showed up, bearing the long-awaited GPS access code (carried by a WHOI staffer who had been authorized for this task by the government) and fresh fruit and vegetables, which were just as welcome. The shuttle also carried a film crew from the Discovery Channel, who came aboard to shoot footage for a TV special on the *Derbyshire* (the film they made was a great success and is still aired periodically).

By April 11, *ARGO* had taken enough pictures and was replaced by *Jason*, accompanied by *Medea*, a vehicle pairing still in use today. *Jason*, a true ROV, is equipped with seven thrusters to propel it, steer it, and control its attitude. *Medea*, a tethered sled, serves as a "garage" for *Jason*. Towed to a position close to the search site, *Medea* absorbs the shocks from the main tether cable while *Jason* swims freely on a 330-foot tether, taking pictures or performing other tasks. *Medea* is also equipped with spotlights to illuminate *Jason*'s work area and a camera that allows *Jason*'s pilot to keep an eye on the tether and prevent it from getting snagged.

On the *Derbyshire* expedition, bad luck struck again: just before midnight on April 11, the telemetry conked out, leaving *Jason-Medea* helpless. The winch went into action, reeling in thousands of feet of cable. It turned out that *Medea* had become thoroughly wrapped in the tether (about a dozen times), breaking the vital optical fibers in its core. Fortunately, the *Thompson* had a spare tether on board.

With its new tether in place, *Jason* was piloted to the wreckage and showed its versatility by using a scrubbing-brush attachment to clean rust and sediment off the broken edges of the metal so that the controller could take high-definition photos of the fractures. It was a slow process because *Jason* had to wait for the fine particles stirred up by the brush to settle before it could use its cameras. From these images, the assessors would be able to deduce what had caused the fractures. The stern of the *Derbyshire* was a particularly hazardous area for *Jason* because of the plethora of jagged edges and protrusions on which the tether could get irretrievably snagged. Here, though, the expedition's luck, helped by skillful piloting, held.

Towing *Jason* and *Medea* between work sites, the crew in the control van were treated to a good view of deep-sea life through the vehicles' cameras. They saw transparent jellyfish with bright-red digestive organs; a strange kind of octopus they dubbed "Dumbo" because of its earlike appendages that reminded them of Disney's cartoon elephant; sea cucumbers in bizarre shapes, including one that resembled "a worm in a baggie"; a creature that resembled a giant centipede; shrimps with extra-long antennae; and galatheid crabs, which Howitt compared to white tarantulas with tails.

The mission encountered yet more problems with stormy weather and a resulting broken tether. Conditions continued to worsen, and a meteorological report indicated that a typhoon was approaching the wreck site. The *Thompson* fled to the safety of some isolated islands far off the southern coast of Japan and waited out the storm.

By April 23, the system had cleared, and the *Thompson* was back on its station. *Jason* and *Medea* were launched the next day, and *Jason* took more images. On April 26, after a final shoot around the stern, the photo documentation was complete. As before, *Jason* had brushed sediment and rust off the corroded steel of the hull before taking its high-definition photos. Getting so close to the ruined stern was the most risky task of all given the numerous jagged edges that could sever *Jason*'s tether. It was therefore saved for last so that the expedition's "eyes" would not be lost before the job was substantially completed.

Jason's work was not over, however. The assessors wanted to collect samples of the hull's metal for analysis in a shoreside laboratory, so the ROV was fitted with a mechanical hole-cutter, a tool resembling a hole saw. But the attachment kept binding and stalling. The backup plan was to use *Jason*'s mechanical claw to pick up small pieces of the wreckage from the ocean floor and then to manipulate a corer to collect samples of the sea bottom. The first attempt to execute the plan failed, however; on the way to the bottom, the computer-connected compass on *Jason*'s bow, which was intended to furnish a heading for the camera to photograph, failed. The crew replaced it with an ordinary compass, and work continued until late afternoon.

When *Jason* finally returned to the deck, the expedition's members celebrated the end of the mission with a group photograph on the stern of the *Thompson*. Controllers sent coded signals to the last of the transponders, which released their anchors and floated to the surface for recovery. Then there was a brief religious ceremony for the dead crew of the ill-fated *Derbyshire*; the captain of the *Thompson* threw a wreath and a memorial plaque into the water. Finally, the mother ship set course for Yokohama, Japan.

Though the evidence was gathered, the work was far from over. Months of testing, with both computer simulations and actual physical models in test tanks, remained. Huge freighters built to scale sailed into a variety of wind and wave conditions on computer screens and in tanks and had their disastrous fates duly recorded. Metallurgists and engineers pored over the photographs of the defunct bulk carrier's fractured metal. Various hypotheses were tested and disproved in labs located in the United Kingdom and the Netherlands. Scientists from Woods Hole took part in the analysis.

A crucial question was whether the *Derbyshire* had in fact cracked and broken in two at Frame 65. The images taken on the sea floor showed conclusively that the hull had not fractured until the ship was many feet below the surface. Ships routinely break up while sinking, as their metal is subjected to significant stresses in directions that the design was not meant to resist. This result proved the shipyard was not to blame for the accident.

Rigorous logic finally defined the fatal scenario. Mosaic images from the sea-floor survey revealed that the ventilator tubes at the bow were missing their protective "mushroom" caps (so named for their shape), which were apparently knocked off by violent waves. Each time a wave crashed over the bow, more water leaked through the tubes into the bosun's storage compartment below. The ventilator tubes measured 20 inches in diameter—minuscule in comparison to the vast interior spaces of the *Derbyshire*—but they let enough water through to gradually make the huge ship bow-heavy.

It is doubtful that Captain Underhill, the master of the vessel, could have detected the loss of the ventilator caps from his bridge, which was located far back at the stern, especially with his vision obscured by the violent storm. Even if he had, he would not have risked sending one of his crew to check; the deck was being scoured by angry waves three stories high—traversing it would have meant certain death.

Gradually, the *Derbyshire*'s bow became so heavy that it could not rise to meet the oncoming waves. Instead, it drove straight into them so that they broke directly onto the deck. An especially powerful wave likely stove in the No. 1 hatch. As the forward cargo hold filled with the sea, the ship lost too much buoyancy to recover. The No. 2 hatch gave way, then No. 3. At that point, the doomed *Derbyshire* was on its way to the bottom. One after another, the remaining hatches came under the direct attack of the storm waves and yielded to the battering. The hull of the *Derbyshire* imploded as it sank, the pieces scattering far and wide. The whole episode was over in a matter of minutes.

During the investigation, one particular question was naturally raised: why didn't the master of the vessel try to avoid the typhoon? A panel of three master mariners concluded that Captain Underhill must have thought initially that he could outrun the storm. After all, his weather forecasting service had predicted it would veer away from his ship's track, to the northwest. Instead, the storm doubled back eastward, heading directly for the lumbering *Derbyshire.*

As chronicled in the dry, terse language of the final report, on September 3, 1980, Captain Underhill was advised of a tropical storm 100 miles to the east of his course. He increased his speed to 12.5 knots—probably all that his huge ship could handle—to get clear of it. On September 4, his charterer's weather service, Oceanroutes, recommended that he turn north. On September 6, the master radioed Oceanroutes that he was resuming his rhumb line course (a path that maintains a fixed compass direction with no deviations) for Kawasaki and reducing speed to 10 knots. But a new danger now loomed for the *Derbyshire*: a tropical depression some 650 miles to her east, which quickly developed into Typhoon Orchid. But the captain relied on his weather reports, which apparently did not indicate that the distant storm was a threat.

At the same time, the captain was also receiving radio weather reports from both the U.S. Navy/Air Force Joint Typhoon Warning Center (JTWC) via Guam and the Japanese National Weather Forecasting Service from Tokyo. Both services substantially miscalculated the track and the force of the storm. Guam projected the storm track to be curving north and west on September 8 and 9, while Tokyo said the track would be much farther to the west. The Guam forecasts also badly underestimated the wind speeds that preceded the center of the typhoon (for Orchid had developed into a full-blown typhoon by September 7). Complicating matters, Orchid had an abnormally large wind field, with consequent high swells running ahead of it. Also, the storm was unusually long lasting, permitting it to further whip up already high waves.

The reports from the JTWC did not anticipate that the *Derbyshire* would encounter 30-knot winds a full 400 miles from the storm's center or 50-knot winds 200 miles from it. Oceanroutes sent another message to the *Derbyshire* on September 8, giving much the same forecast as Guam, but never advised the master to alter his course. In effect, Captain Underhill was flying blind.

On September 8, the *Derbyshire* ran into such heavy seas that the master reduced speed to 2 knots—the absolute minimum required for maintaining steerage way. It was too late to cut and run. Turning away from the center of the typhoon would have left the vessel broadside to the waves until the turn was nearly completed—and a huge ship like the *Derbyshire* takes a long time to turn under the best conditions. The risk of capsizing under the circumstances was too great to take. Even if he made the turn successfully, the captain would have risked being swamped by waves coming from astern.

With escape out of the question, Captain Underhill's only choice was to keep the vessel headed into the wind, with the seas coming in from 10 to 45 degrees off the starboard bow. Centuries of experience have shown that this is the safest position for a ship in relation to approaching waves. It is, in fact, the course recommended in the Royal Navy's *Mariner's Handbook NP-100.*

A typhoon is a low-pressure zone; north of the equator, the winds spiral in toward its center counterclockwise. For a ship at sea, the most perilous part of a typhoon is the so-called dangerous semicircle, or the northeastern half of a line passing through the center of the eye. It was in that treacherous spot that the *Derbyshire* found herself on September 8, with winds now averaging 46 knots—violent enough to blow a person away. The significant wave height, or *average* height of the highest one-third of the waves, was better than 26 feet (which meant some waves were even higher). A ship some 80 miles from the *Derbyshire*, the *Alrai*, reported wind speeds of force 12—over 70 miles per hour—and waves 60 to 100 feet in height. A ship in the midst of such waves would be pitching and rolling and yawing uncontrollably. The stresses on its framework would be unbelievably severe. We cannot know the height of the waves that knocked off the ventilator caps at the *Derbyshire*'s bow nor the height of those that stove in her hatches—but they must have been immense. Even so, the experts of the "Re-Opened Formal Investigation" agreed that, if the *Derbyshire* had not become bow-heavy, she probably would have survived the typhoon, although with damage to her superstructure. Unfortunately for the crew and their families, the *Derbyshire* met a very different end.

The report ended with recommendations for improving the safety of bulk carriers like the *Derbyshire*. They included a more wave-resistant design for ventilator caps, stronger cargo hatches, and an electronic system to monitor the depth of water in the forward hull spaces. The assessors also called for forward pumps that could be operated from the bridge at the stern of the vessel and a voyage data recorder (VDR), the maritime equivalent of an aircraft's black box.

* * *

Tragically for the *Gaul*, a British trawler that vanished without explanation in a bitter Arctic blizzard some 55 miles north of the North Cape of Norway on February 8, 1974, none of the safety devices for megacarriers recommended in the final report on the *Derbyshire* disaster was available.

The *Gaul* was a fairly new ship, only two years old at the time she went down with all thirty-six of her crew. Two hundred and sixty-three feet in length, she was considered a "supertrawler," fit for lengthy journeys into stormy, ice-choked waters. As the 375th ship off the line at Brooke Marine Ltd. Shipyard, she had the benefit of experienced architects and engineers behind her. Her frames were more closely spaced than is customary to give her hull greater strength. Her powerful engine drove a propeller whose four blades had adjustable pitch, like the propeller blades of pre-jet aircraft. The propeller was fitted with a Kort nozzle; this housing enhances the propeller's force by increasing the volume and flow of water through a sort of jet effect. The propeller was also capable of turning from side to side, resulting in very sensitive steering.

Based in the east-coast port of Hull in Yorkshire, England, the *Gaul* was classed as a beam trawler, meaning her nets were fastened to a horizontal beam that was dragged through the water behind her. She also carried an onboard plant for filleting and freezing fish, making her the model of a self-contained operation.

At 9:30 A.M. on that fateful day in February, this sleek, modern, blue-and-white trawler had a radio conversation with another British trawler about a mile away. The mates of the two vessels discussed weather conditions at nearby fishing grounds. Then the mate of the *Gaul* said, "We're going to dodge more in to land." Later the *Gaul*'s radio operator sent two personal messages from crew members to the U.K. The last message was sent at 11:09 A.M. Nothing more was heard from the *Gaul*. Around the same time, other trawlers in the area reported that severe blizzard conditions were forcing them to leave their fishing stations and seek shelter. The *Gaul* never returned to port.

The Norwegian Rescue Coordination Center at Bodø instituted a search-and-rescue mission on the afternoon of February 11, as soon as the storm had subsided enough to permit operations at sea. A British aircraft carrier and naval oil tanker, in the area for a joint exercise with the Norwegians, took part in the search. Together, ships and aircraft covered an area of 177,000 square miles. Twenty other trawlers in the area helped in the hunt for their missing comrade. After four days of intensive searching, not the slightest trace of the missing *Gaul* was found—no bodies, no life rafts, no wreckage, no patches of oil on the water's surface. Intermittent searches continued for the next couple of months, but nothing turned up.

The Marine Accident Investigation Branch (MAIB), with a few sketchy facts to work with, concluded that the *Gaul*, like many other unfortunate fishing vessels, had been overwhelmed by storm waves and foundered. But many thought there was more to the story. The families of the lost crew suspected that the authorities were holding back information that they did not want the public to know.

For years before the *Gaul*'s mysterious disappearance, rumors had circulated among Britishers that their government used the crews of fishing boats to collect intelligence on the Soviet Navy. In the wake of the tragedy, the press ran embarrassing questions related to the *Gaul*'s activities. Had the crew been trawling for more than fish? Did the Russians seize the ship? If so, what did they do with the crew? (The Russians had a well-earned reputation for imprisoning and torturing foreigners who had run afoul of them.) Or had they shelled or torpedoed the *Gaul*? At the time, the Barents Sea was fairly crawling with Soviet and NATO warships, nuclear submarines, and surveillance vessels, each trying to spy on the others. And the great Soviet naval base at Murmansk lay nearby. Meanwhile, the British government kept obstinately mum on the subject.

On the morning of May 8, a single clue materialized: a life buoy bearing the words *Gaul* and *Hull*. It was discovered by the captain of a small Norwegian whaling ship about 15 miles off the northern coast of Norway and about 20 miles southeast of the *Gaul*'s last reported position. The captain took it that same day to the nearest shoreside police station, in the port of Vardø. He left the ring's coating of algae intact, figuring that the organisms might yield information about the temperature and salinity of the water in which the life buoy had been found. Those facts could then be correlated to oceanographic surveys of the Barents Sea and bring investigators closer to finding the *Gaul*. A few days later, the life buoy was back in Britain being scrutinized by a forensics team.

But the life buoy and its load of algae yielded little information. Materials experts sliced into the foam plastic that gave the ring its buoyancy and concluded that the extent to which it was crushed by water pressure showed it had never been deeper than about 65 feet below the surface and had probably drifted either on the surface itself or no more than 33 feet below it. The algae, various species of diatoms, were equally enigmatic. About two-thirds of them were types that grow in shallow seawater along rocky coasts; the remainder belonged to species that thrive in brackish or even fresh water. Investigators concluded that the life buoy had been either carried close to the coast by winds and currents and stayed there or discovered by the Russians, who later planted it at sea in a new location to confuse the search.

A few tantalizing clues popped up later in the month. Each of two trawlers pulled up a chair of British make, but it turned out that neither was from the *Gaul.* Another trawler, from the *Gaul*'s home port of Hull, snagged its nets on a previously unknown underwater obstruction about 55 miles southeast of the *Gaul*'s last reported position and later in the day pulled up another fishing vessel's lost net. The British Navy sent an anti-submarine frigate to investigate, but its sonars found nothing after searching 36 square miles of sea floor. A West German fisheries protection ship in the area reported that two of its trawlers had snagged their gear on an unknown obstacle about 50 miles east of the *Gaul*'s last known position; one of them had pulled up pieces of a guard rail, but these were overgrown with seaweed, indicating that they had been on the bottom long before the *Gaul* disappeared. Then yet another trawler from Hull twice picked up sonar reflections from a large obstruction on the sea floor about 100 miles east-southeast of the *Gaul*'s last known position. The marine office in Hull examined the sonargram, but the results were inconclusive.

The ensuing months brought more enticing leads. The most solid was a report from a Norwegian trawler, the *Rairo,* which on November 15, 1975, was trawling in a spot about a half hour's sail from the *Gaul*'s last known position. Its echo sounder (a type of sonar used to locate schools of fish) picked up a sharp rise on the bottom; the captain swerved sharply to avoid it. The device showed a trace of some large object on the sea floor. When the *Rairo* pulled in its net shortly thereafter, the crew found almost all of the trawling gear from another vessel. The crew turned over the lost net and the sonar trace to the British authorities, who determined that the net did not come from the *Gaul*.

Another piece of the puzzle, a yellow plastic writing case or wallet, washed onto a beach deep inside a fjord east of Norway's North Cape. In it were two names and addresses, one of which belonged to a sister of a *Gaul* crewman named Colin Naulls. A Norwegian couple had made the discovery in March 1974, but British authorities did not hear of the find until early 1977. In the end, they concluded the wallet was no more helpful than the life buoy in tracking down the *Gaul*'s resting place since it could have accidentally fallen overboard before the tragedy occurred.

Then, in March 1977, the British trawler *Marbella* hauled in its nets and discovered half of a container for an inflatable life raft. The captain radioed the serial number on this piece of wreckage back to Hull, where the *Gaul*'s owners confirmed that it did indeed belong to the missing ship. A few months later, the *Ella Hewitt*, another trawler, found a green plastic sheet among the fish in its net; it bore the words *GAUL BOATS COVER*. This lifeboat cover was picked up about 25 miles northwest of the *Gaul*'s last reported position. But the British authorities decided that, in the three and a half years since the *Gaul*'s sinking, bottom currents had had plenty of time to waft the lifeboat cover to where it was found.

A few more reports about an unidentified object on the sea floor in approximately the same spot as the *Rairo*'s discovery trickled in from Norway. The British officialdom was finally confident that this was the region in which the *Gaul*'s remains lay, but the government took no action, maintaining that the cost of an underwater investigation was too great.

A false lead that caused great anguish in Britain was the discovery of unidentified bodies that washed up on a beach on Russia's Barents Sea coast, carried there by the last faint currents of the Gulf Stream. Russian border guards found one body on April 11, 1974. It was immediately investigated by the KGB, the Soviet secret police. After an autopsy, it was buried in a graveyard in a small town on the Soviet-Norwegian border. A second body, found only eleven days later, was treated less ceremoniously: it was autopsied on the beach where it was found and buried there after its bones had been removed and sent to the Russian naval base in Murmansk for tests. On the last day of April, a third body was found. It was heavily tattooed, with designs and mottoes that the Soviet authorities concluded were English. This body was examined and photographed; the official Soviet reports did not say where it was buried. The type of clothing on these corpses and the fillings in the teeth of one of them led the KGB investigators to believe that they were English, possibly crew from the *Gaul*. However, DNA comparisons between the cadavers and family members of the *Gaul*'s crew later proved that there was no connection.

The big break in the case came in August 1997 when British TV producer Norman Fenton chartered a survey vessel and located the long-sunken trawler at 72°04.1' N, 25°05.3' E. It was lying about 920 feet down, canted over on its starboard side at an angle of about 35 degrees. Mr. Fenton had done his homework well. He tracked down a number of the reports from trawlers that had lost their gear on bottom obstructions—presumably wrecks—in the area where the *Gaul* had disappeared and narrowed his search site down to a "box" of about 2 miles × 1 mile. He began a search of this practicable area with side-scan sonar. After some five hours of exploring with sonar, Fenton's crew spotted a wreck approximately the same size as the *Gaul*. The ship performed additional passes to obtain a profile of the wreck, which again matched that of the *Gaul*. The surveyors dropped anchor close to the wreck and deployed an ROV.

The ROV reached the sunken wreck amidships on the starboard side. Its cameras sent back images of a fishing net draped over the starboard funnel (the *Gaul* had a funnel on each side of its hull). More lost trawl nets covered much of the ship. Powering itself forward, the ROV found the names *Gaul* and *Ranger Castor* (the *Gaul*'s original name) on the bow, still legible. Another maneuver brought the ROV and its electronic eyes to the bridge, where a shipyard plate provided more identification. A telescoping arm thrust a camera into the ship's interior to look for bodies and transmitted images of objects that vaguely resembled parts of human skeletons. However, a British forensic pathologist found the images too indistinct to make any positive identification.

Fenton's film was a sensation. The government was forced to reopen the formal investigation it had abandoned in 1974. In 1998, the MAIB had no choice but to schedule a full-scale investigation of the wreck of the *Gaul* for August—past the time when weather over the Barents Sea would be at its most moderate. It chartered a Norwegian survey ship from the North Sea oil fields, the *Mansal 18*, a vessel so large that it sported a helicopter landing pad on its bow. The expedition set out with several ROVs, three representatives from the *Gaul* Families Association, and a TV crew from the BBC. Reporters from the *Hull Daily Mail*, a paper that had run a long series of articles on the *Gaul* and the British government's covert use of trawlers as spy ships, were pointedly excluded. (The defiant newspaper sent a reporter to Norway anyhow, to report on developments from the port of Tromsø.)

Using the coordinates from the Fenton expedition, the *Mansal 18* had no trouble locating the *Gaul*. Following standard procedure, the crew used side-scan sonar to zero in on the target, then sent down camera-equipped ROVs to document the wreck. The plan also allowed for the use of small explosive charges to blow open the *Gaul*'s portholes so that the cameras, on their extendable arms, could penetrate the crew's living and recreation areas to search for human remains. It was a high-tech operation, and the results were enough to prompt the MAIB to reopen the formal investigation.

The MAIB's video images—forty-five hours' worth—showed some significant evidence. A deep dent in the port funnel indicated that it had suffered a heavy blow, as if from a huge and violent wave. The propeller was set slightly to port, indicating the *Gaul*'s course; the pitch of its adjustable blades was estimated at 50 to 75 percent full forward speed. The factory companionway door, which led from the trawl deck at the stern to the filleting and freezing section, was wide open; a net had been sucked inside it. The engine-room escape door, normally kept closed, was also open, as were two very heavy hatches used for loading fish as they came out of the net. The video images showed that the hatches were locked open by the ship's hydraulic system. This discovery led the MAIB investigators to wonder whether the *Gaul* had actually been fishing during this violent storm, but the videos showed that the port trawl board (one of two boards on either side of the net that keep it open) was securely stowed on deck and that the trawl's wire cables had not been used. Conclusion: the *Gaul* had not been fishing when it sank.

Finally, there were mysterious indentations on either side of the bow. Had the *Gaul* been rammed by a Russian ship? Did it collide innocently with another vessel in the blinding snow? No, for a collision would have bashed in only one side of the bow. So what caused these massive dents? And why did the rest of the bow show virtually no damage? There were no signs of fire (e.g., blackening of the engine-room walls), explosion, or military damage.

John Moore, a Briton who saw the published photos, believed they revealed a 4-foot hole in the *Gaul*'s bow located 7 feet above the waterline, proving that the *Gaul* had been holed by a "massive submerged object." His opinion carried a good deal of weight since he was the retired manager of the shipyard where the *Gaul* had been built. In fact, he had been in charge of the construction of the *Gaul* and her three sister trawlers. He had even sailed on the *Gaul* during her first fishing trip and found her "a marvellous ship, one of the best we ever built." Moore went on to explain how the *Gaul*'s hull had been specially reinforced to withstand the violent waters of the Barents Sea and its floating ice.

Moore's opinion notwithstanding, the MAIB used its now considerable data to extrapolate the following scenario. The *Gaul* had been steering south toward shelter, with the wind and waves coming at an angle of 50 to 55 degrees off the port bow. The 263-foot ship could handle such conditions, but, unexpectedly, a giant wave slammed into her port side (accounting for the dented port funnel) and knocked her around so that she lay parallel to the waves—the most dangerous possible position for a ship in a storm. More giant waves assailed her, knocking her over on her starboard side, where she lay at an angle probably more than 90 degrees. Anything loose inside the hull, including tons of frozen fish, obeyed the law of gravity and tumbled into the starboard side, preventing the ship from righting herself as she had been designed to do. Video images of the ship's interior showed the way the cargo had shifted.

Doors and hatchways flew open under the force of the ship's motion. The rear deck, where the trawls operated, flooded because mats and other debris clogged its drainage ports. Water poured in, flooding the hull. The ship sank stern-first, probably within ten minutes. The air trapped inside the hull rose into the bow, filling it. When the ship had sunk about 100 to 130 feet, the water pressure overcame the pressure of the trapped air and pushed the bows in, accounting for the dents, which measured about 6 feet × 1 foot.

The MAIB built a scale model of the *Gaul* to test this hypothesis. Trials in the testing tank showed that waves the size of those the *Gaul* encountered could have rolled the ship over past 90 degrees in a mere seven seconds. When this happened, the fish-loading hatches would open of their own weight if not tightly dogged down. The rest followed logically. But there was still the question of the distress call—why was none sent? The answer remained elusive, but the MAIB suggested it would have been extremely difficult to accomplish anything with the ship lying on its beam ends. And why did none of the crew escape when they had ten minutes to flee the sinking ship? Again the *Gaul*'s extreme tilt would have made it virtually impossible for anyone to move, much less launch a life raft. And the chances of survival in such huge waves and Arctic temperatures would have been vanishingly small anyway.

The MAIB made a number of recommendations in its final report. First among them was the mandatory equipping of all vessels over 100 gross tons' displacement with a VDR. Another was requiring all ships to report their whereabouts each day. Ironically, the *Gaul*'s owners had made these reports a part of their standard operating procedure. Unfortunately, on the day of the tragedy, the *Gaul*'s captain had not updated his position. (This recommendation met with resistance from commercial fishermen, who did not want to reveal daily the location of their favorite fishing grounds.)

The MAIB also recommended that all vessels carry an emergency position indicating radio beacon (EPIRB). This device is supposed to pop loose as soon as it has been submerged to about 10 feet, shoot to the surface, and start broadcasting a distress call, which is relayed by satellite and guides rescuers to the site. However, EPIRBs are often poorly maintained. As a result, they do not release themselves as intended from a foundering vessel. Or their batteries are allowed to run down so that no radio signal is sent even if the device does surface. Still another recommendation was for shipboard transponders that would automatically indicate location via satellites. A more modest suggestion was to equip all vessels with an attached sonar beacon that could ping out a wreck's location from the sea floor. (Since 2000, all European fishing vessels over 77 feet in length have been required to carry equipment capable of transmitting their locations to British fisheries management centers. While the requirement's primary purpose is to keep tabs on boats and prevent the vessels of one nation from trespassing in another nation's waters, the signals can also be used to locate wrecks more quickly and accurately.)

Since some fishing vessels operate beyond the effective range of their radio transmitters, the MAIB suggested that all vessels' radios should be upgraded (if necessary) to a certain standard. Many experts agreed wholeheartedly with this particular recommendation, but, curiously, most captains opposed it—perhaps because of the added cost.

One of the final MAIB recommendations directly addressed public dissatisfaction with the government's handling of the *Gaul* tragedy. For years, officials had shrugged off requests for information from the dead crew's families and kept many relevant facts about the incident secret. When the United Kingdom passed a freedom of information law in 1987, much long-buried information came to light. But in the case of the *Gaul*, the Ministry of Defence (MoD) had destroyed all records. MoD explained that it was a long-standing policy to destroy any papers more than five years old unless they contained material of historical or administrative value. Some information regarding the *Gaul* survived in the archives of other government departments, but it was spotty. The public was incensed. In response, the MAIB suggested strongly that the government should do everything in its power to keep relatives of lost seafarers informed and to do so in a sympathetic manner.

Not until 1997 did the government come clean about its use of fishing vessels and their crews to gather intelligence. One retired sea captain, Jack Lilley, recalled a James Bond–like experience from 1968. He was scheduled to take his trawler to Leningrad for an international exhibition of fishing vessels. Shortly before he left, a naval commander came to Hull from London and showed him aerial photos of the Leningrad shipyard, telling Lilley that the exhibition offered a unique opportunity for gathering information. He gave the captain a camera and told him to take as many pictures as possible of designated docks and warehouses. These were the sheds in which the Soviets were building their newest nuclear submarines.

The naval commander added that, if the trawler captain were caught, he would be on his own—his government would deny having used him and would give him no help. Almost as an afterthought, Lilley was asked to photograph the major Soviet naval yard at Kronshtadt as he sailed by on the way into Leningrad.

A week after tying up in Leningrad, Captain Lilley took photographs of the submarine sheds without being caught. But when he started back to England, he was tailed by a series of Soviet gunboats, whose crews ordered him to stop and be searched. As Lilley was still within Soviet territorial waters, he felt it prudent to obey. He felt it even more prudent to stuff the camera and exposed film in a bag and ditch them over the side of his ship. "I dread to think what would have happened to me and the lads if they had found the camera on the boat," he told the *Hull Daily Mail.* Though British officials had finally admitted to having used fishing boats from Hull and other ports to gather information on the Soviet Navy and its operations, they refused to verify Captain Lilley's tale, along with a number of other reports from fishing captains.

In 2000, a respected officer from the Royal Navy's Defence Intelligence Staff made a rather disingenuous statement about the *Gaul*'s possible involvement in such intelligence-gathering activities, saying that no record in the MoD files indicated that the *Gaul* or any of its crew had engaged in intelligence operations for the Royal Navy. In rebuttal, the lawyer for the *Gaul* Families Association caustically asked about records that the intelligence officer had *not* examined or that no longer existed. The lawyer also raised the possibility that the *Gaul* was spying on behalf of another agency, such as the fabled MI6.

With these questions still lingering, the MAIB's final report turned out not to be final after all. As the public continued to demand answers, the MAIB mounted another expedition in the summer of 2002. Prominent among its stated objectives were searching for human remains and discovering if the *Gaul* was carrying intelligence-gathering equipment at the time she sank. The MAIB also hoped to get fresh data that would help to determine the cause of the sinking.

The MAIB hired a huge state-of-the-art ship, the *Seisranger*, normally used for oil-field exploration and owned by a subsidiary of Halliburton, the transnational corporation formerly headed by U.S. Vice President Dick Cheney. For this trip, the *Seisranger* carried nine ROVs: two large work-class vehicles, two medium-sized observation-class vehicles, and five mini-ROVs that were designed to explore inside the *Gaul*'s hull. The mini-ROVs would ride beneath the larger ROVs until they reached the work site; there they would be piloted off on their own—much like the *Jason-Medea* combination used so successfully on the *Derbyshire* expedition. Rounding out the armory of search equipment were a side-scan sonar and a multibeam sonar array called the *Sea Bat*, designed to transmit images of very high resolution.

The *Seisranger* departed from Aberdeen on June 24 and stopped at Bergen, on the west coast of Norway, to refuel. It took the next two days to travel up the long Norwegian coast, round the country's northern tip, and reach the wreck site, some 70 miles from land. The team's first priorities were to calibrate the sonars and other equipment and place a sonar beacon on the sea floor. This task accomplished, an ROV went down to get an accurate fix on the location of the wreck and check for fishing nets. (Nets around a wreck are always a hazard for ROVs, which can easily get entangled in them.) The investigators also performed a side-scan sonar survey of the sea floor around the wreck.

The area's good-weather season was short, so the crew worked fast. *Seisranger* arrived on site at 6:10 A.M. on July 2; an ROV was in the water within an hour; and by 8:48 A.M. the *Gaul* was officially located, and the ROV was sending back good-quality images.

The following day was a big day for the sonar. The *Sea Bat* completed its survey, and the photo mosaic of the sides of the hull was completed. In the control van, the forensics team analyzed the results of the side-scan sonar survey. Meanwhile, the hard-working ROVs blasted the *Gaul*'s port funnel and the Kort nozzle partly clear of sediment and fouling organisms with powerful water jets. An ROV also made a preliminary reconnaissance of the doorway that gave access to the factory deck where the fish were cleaned and frozen.

On July 4, the ROVs searched the sea floor for the *Gaul*'s starboard trawl wire and trawl board. They found a number of trawl wires from other fishing boats, and the daily report from the MAIB noted tersely, "The quantity of fishing nets and trawl gear on the wreck has increased since the 1998 survey." All or most of this mass of extraneous materials had to be cleared away so that the ROVs would not become entangled in it. During the cleanup, one of the ROVs did become snared in a knot of fishing gear on the *Gaul*'s deck and had to be cut free by another ROV.

While the ROVs removed debris from the wreck, technicians on the *Seisranger* prepared tools that would measure the indentations on the *Gaul*'s port funnel and the angle of its rudder; the latter could potentially provide a clue to the captain's last maneuver.

By the end of July 11, the angle of the rudder had been measured, and one of the ROVs was fitted with a diamond-impregnated wire saw to cut it off so that a camera could photograph the pitch of the propeller blades. Photogrammetric analysis would enable the forensics team to calculate the exact angle of the blades and determine the speed at which the *Gaul* had been traveling. Another ROV used its manipulator to smash in some of the wreck's portholes and fit them with sleeves through which a spyball camera could be inserted. The spyball recorded interior shots from a number of compartments—and then it got snagged and was lost. A Norwegian Coast Guard vessel delivered a replacement. The Norwegians also brought an engineer to repair a mini-ROV with mechanical problems and a steel-cutting specialist to help the team cut into the *Gaul*'s hull.

On succeeding days, a *VideoRay* mini-ROV, an American product specially designed for this expedition in cooperation with the Canadian firm Inuktun, went below with a work-class ROV fitted with the replacement spyball camera. Together they photographed the radio room, the chart room, the hatch used for unloading fish, and several other interior spaces. Simultaneously, other ROVs worked to clear the silt covering the propeller's lowest blade so it, too, could be photographed. The investigators took a close look at the root of the blades to see if anything there had gone amiss. Still other remotely controlled vehicles, finished with clearing debris from the ship's deck, worked on removing a great tangle of nets and wire cables from the sea floor so that the investigators could search for the starboard trawl board. The *Seisranger*'s crane was called into action to lift them out of the way with a cable whose lower end was attached to the nets by an ROV. At one point, the weight of this sea-floor junk was so great that it made the massive *Seisranger* heel slightly.

On July 16, an ROV pilot flew the *VideoRay* through one of the fish-loading hatches and the fish chute into the processing area—"A hazardous route," commented the daily report. The little *VideoRay* got some excellent video footage of the processing area and of the radio room, chart room, and wheelhouse. More windows had to be broken to give the ROV access. The starboard trawl board was found when the *Seisranger*'s crane moved a heap of netting on the seabed, and an ROV photographed it *in situ*.

The following day, a camera recorded images of the crew's mess room and its passageway, and an ROV began cutting away the wooden door that led to the officers' cabins. The ship's crane pulled "a very large quantity" of fishing nets off the forward part of the wreck to free up more portholes. These portholes were smashed to provide further access to the officers' mess, for the search had to be as thorough as possible to avoid future accusations of neglect. Even as this work proceeded, another team was trying to cut an opening into the steel walls of the hull. They encountered difficulties but eventually cut a 22-inch hole into the port side of the wreck, creating ample room for the *VideoRay* to enter.

The investigation was running behind schedule due to earlier mechanical problems, so the search team decided to cut a 60-inch hole into the hull, allowing one of the medium-sized ROVs to gain entry and speed up the exploration of the *Gaul*'s interior. A blast of grit from a jet device mounted on an ROV easily cut through the hull plating, but it could not get through the heavy steel framework behind it. The investigators resorted to emergency measures and flew in a demolition expert that very evening. Shaped charges would complete the job.

But the experts changed their minds overnight, reasoning that explosive charges would do too much damage to the interior of the *Gaul.* They compromised by using the grit blast to cut another aperture. The sea, which had been more or less calm for most of the expedition, suddenly developed a 10-foot swell, making it very hard for the ROV operator to hold his cutting tool in position. He completed his task nonetheless. The MAIB's daily report noted that, between them, the *VideoRay* and the big ROV fitted with the spyball camera had surveyed over 90 percent of the lost trawler's living areas. The vehicles had examined seven compartments in twenty-four hours and had even been pulled back for repeated looks at particular objects.

Despite this accomplishment, the mission's prime objective, to find and bring home the remains of at least some of the drowned crew, was not yet complete. British authorities, realizing the political significance of the matter, extended the survey by twenty-four hours. The investigative team worked frantically around the clock. Grit blasts and abrasive disks gnawed away at the existing entry holes, enlarging them enough for the ROVs to "retrieve items of interest"—code for body parts or clothing—"for forensic examination."

Finally, on July 24, the forensic team announced that they had recovered samples of remains from six individuals. The mission was wrapped up by 3 A.M. on July 25. The expedition's leaders held a brief ceremony on the *Seisranger*'s deck to honor the *Gaul*'s dead crew, and the expedition prepared to sail back to Aberdeen, a four-day voyage.

DNA analysis of the recovered bones and the process of matching those results with samples from surviving relatives took time, but in January 2003 the MAIB announced that three of the *Gaul*'s lost crew had been identified. The family members greeted the news with relief—at last they had confirmation of the fates of their loved ones, even though the cause of the tragedy had still not been established.

The *Gaul*'s story is not yet concluded, for the coroner's investigation, originally scheduled for early 2003, had still not begun by mid-July. It is expected that the investigation will focus mainly on the government's actions and motives in keeping matters secret for so long, for no one can truly re-create the grim scene at the moment when the forces of nature overwhelmed the struggling trawler.

* * *

On January 16, 1998, the *Flare*, an elderly bulk carrier of Cypriot registry, went down in the Cabot Strait off the eastern coast of Canada. Unlike the *Derbyshire* and the *Gaul*, the *Flare*'s demise was extremely well documented; two Canadian light planes and a helicopter took photographs of the event, and four of the crew survived to give testimony.

Built in Japan in 1972, the *Flare* measured 580 feet from stem to stern. Her diesel engine drove a single, fixed-pitch propeller at a running speed of 15.1 knots—she was a strictly utilitarian ship. Designed mainly for carrying grain, she had seven dry cargo holds; the No. 4 hold doubled as a water tank for ballast. Constructed of Lloyd's grade A steel, she had a double bottom for extra protection.

This was the *Flare*'s captain's first voyage as master of a bulk carrier, although he had served as first mate on three similar ships. Greek himself, he commanded an international crew: three were Greek nationals; three were Yugoslavs; two were Romanians; and sixteen were Filipinos. Some of the crew spoke—and understood—English very poorly, an impediment to the voyage because English is the designated language of international maritime operations. When the captain wanted to communicate with his Yugoslav radio operator, the chief engineer had to translate his English to Serbo-Croatian.

The *Flare* left Rotterdam, the Netherlands, on December 30, 1997, with her cargo holds empty. She was scheduled to pick up a cargo of grain in Montreal, Canada. She was lightly ballasted, probably to save on fuel. She had some cracks in her deck plates, and the steel framing around some of her cargo hatches needed replacing. Management had decided to save time and wages by having the crew make these repairs at sea rather than in port at Rotterdam. So on her deck the *Flare* carried over a ton of steel plates, flat bars, and a portable welding machine; a welder in the crew was to perform the needed repairs en route to Montreal. As the aging carrier left port, the pilot noticed that she was light in the bow and heavy in the stern. This unorthodox distribution of weight would have serious consequences later on the voyage.

The weather turned nasty almost immediately. For twelve days (from December 30 to January 10) the wind speed varied between force 9 and force 11 (44 to 60 knots, or 50 to 69 miles an hour, just short of hurricane strength). The waves ranged from an estimated 33 to 52 feet—the heights of a three-story and a five-story building, respectively. On January 10, the wind died down to a comparatively moderate 17 to 20 knots, or 20 to 23 miles per hour, and the waves diminished to an estimated 8 feet in height. The entire time, the *Flare*'s lightly ballasted bow skipped over the tops of the waves instead of riding through them. The ship pitched, pounded, and slammed violently on the crests and hollows of the seas. The crew noticed the hull of their ship flexing and bending and twisting beneath their feet. As the survivors later reported, they had trouble sleeping and eating because of the hull's violent gymnastics. The rocky voyage had one of the crew so worried that he slept with his cabin light on and practiced getting into his warm clothes as quickly as possible. On January 14, the wind picked up again, with speeds ranging from 25 to 60 knots, or 29 to 69 miles per hour; the waves ran from 10 to 20 feet.

The already apprehensive crew would have been even more apprehensive had they known their ship's history: the *Flare* had been poorly maintained over the years. Her last two inspections showed severe rusting of the internal framing and bracing, a problem her owners temporarily remedied by sandblasting the rust away and repainting the metal. In 1995, some of the *Flare*'s framing had been replaced while she was in port at Shanghai, a place that offered skilled and cheap Chinese labor.

Most commercial ships are monitored by classification societies, which are industry groups that develop rules for ship construction, oversee the work at the shipyard, and inspect vessels regularly. The *Flare*'s classification society was Lloyd's of London, which inspected her at Cienfuegos, Cuba, on November 28, 1997. Thanks to some hasty repairs, the *Flare* barely squeaked through her inspection, and Lloyd's insisted that she undergo major repairs to be completed before the end of February 1998—just two months from the date she left Rotterdam.

As the *Flare* crept along south of Newfoundland, the waves grew higher as the sea floor shallowed. Wind and current were moving in opposite directions, a condition that produces rogue waves of unusual size and force. Despite the storm, the master took care to change the water in his ballast tanks, as required by Canadian regulations. (Changing the water prevents the accidental importation of exotic animal and plant life, either of which can take over an ecosystem and become an aggressive pest.) He also told his employers' radio operator that he had filled the ballast tanks to make the ship ride lower in the water and keep the propeller out of the way of floating ice.

At about 8:00 A.M. on January 16, the crew were startled by a very loud bang as the bow whipped up and down. The entire hull followed suit; it began to bend up and down and vibrated worse than ever. The ship's general alarm went off. The crew rushed up to the deck from their quarters in the stern only to discover that the *Flare* had broken in two.

The wounded ship carried two sizable lifeboats with motors, but they were of no help. The crew could not even reach the starboard lifeboat because the stern, where they were congregated, was now listing heavily to starboard. And neither it nor the port lifeboat could have been launched because they had been tied down with extra lashings to keep the turbulent seas from sweeping them away. In the dark, and with the deck covered with ice from freezing sea spray, untying them would have been an impossible task.

Some of the crew managed to wrestle a life raft down one deck. They pushed it over the stern and made it fast to the railing with a line. The captain shouted to them not to abandon ship, because the propeller was still turning, and he feared for their safety. Then the line chafed through, and the life raft drifted away. One can only imagine the crew's feelings as they watched their last chance at safety disappear.

The stern sank in about half an hour. Before it did, some of the crew saw the bow of a vessel approaching them. They took it for a rescue ship. But it was the bow section of their own ship; the propeller, still turning, had driven the stern on a course that brought it back near the bow. Just before the stern heeled over further and sank, the radio operator managed to send out a *Mayday* signal, which was picked up by a Canadian Coast Guard station in Newfoundland. The signal was indistinct and incomplete and required precious minutes to interpret.

As the stern sank, all the crew except for the chief engineer, the third engineer, and one other jumped over the side. All were wearing life jackets, so they would at least stay afloat. Somehow the port lifeboat came free—and capsized. Six of the crew managed to swim to the upside-down lifeboat and clamber aboard despite the huge waves that tossed both boat and swimmers about.

The Canadians sent out a call for assistance to all vessels within a 100-mile radius of the *Flare*'s last reported position. Fifteen responded to the call; four were chosen for the mission. A helicopter and two fixed-wing aircraft were also dispatched. One of the planes had sophisticated radar and infrared search equipment, much-needed gear in the dark, which would remain for a few more hours. When the sun finally rose at 11:45 A.M., the searchers had found no trace of the ship or her crew.

At 2:09 P.M., one of the search planes found the bow section of the *Flare*. Shortly thereafter, one of the helicopters spotted a large oil slick about five miles southwest of the drifting bow, and at 2:23 P.M. it saw the capsized lifeboat with four sailors clinging to it. Their two shipmates had weakened from the piercing cold and lost their grip. It took only eleven minutes to lift the near-dead survivors into the helicopter, which took them to a hospital on the French island of Saint-Pierre off the Canadian coast.

Two other helicopters proceeded to the scene. They saw the bow and a huge oil slick beyond it, spreading over an area of sea measuring 10 miles × 3 miles. They also found two capsized lifeboats—apparently the starboard lifeboat had also come free—and two life rafts. Search-and-rescue (SAR) team members were lowered to one of the life rafts. It was covered with oil, but there were no survivors aboard. The SAR team punctured and sank the life raft to prevent a duplicate effort by other search craft.

Other rescuers were lowered to the lifeboat from which the four survivors had been pulled. Without SCUBA gear, one swam under the boat. There he found a body so entangled in ropes that he could not free it. (The body was later retrieved by a Coast Guard ship.)

Floating in the oil-covered water were thirteen bodies; the SAR personnel recovered only eight (four in each helicopter) because the fumes from the oil slick made them nauseated and the oil made the task a dangerous one. The remaining bodies were pulled up by Coast Guard vessels. The dead crew were all wearing life jackets, but they were lightly clothed—most had apparently been too stunned to put on socks and shoes. All were covered with a slippery layer of fuel oil. By 9 P.M., fourteen bodies had been recovered. Seven were missing and would never be found.

The *Flare* had been carrying immersion suits, which can prolong survival time to between twelve and fourteen hours in water just above freezing temperatures, but the crew had not been able to don them. Besides, there were only six immersion suits for a crew of twenty-five men. The lifeboats contained enough "thermal protection aids" for the whole crew, but the men did not know where they were stowed in the boats and, as mentioned earlier, could not have reached them in any case.

The broken-off bow floated about for four days, finally sinking southeast of the eastern tip of Nova Scotia. While it floated, investigators took aerial photos and videotape of the damage. Video was also taken from the deck of one of the freighters called to aid in the search. The images showed that the metal of the hull had become embrittled through a combination of repeated flexing and low temperatures—it could have snapped in two at any moment.

A major contributing factor to the disaster was badly distributed ballasting. Throughout the voyage, the *Flare* rode with her bow high in the water, making her bounce and flex. The bow trim was consistently shallower than both her loading manual and Lloyd's rules specified. No one knows why the master did not follow instructions or why he told the ship's owners that the ballast tanks were full when they were not. The patches of snow and ice on the deck of the bow showed where the seawater ballast in the tanks was located. Where the tanks were full, the latent heat of the water (the heat given off as the water freezes) had melted the snow and ice away. Where the tanks were less than full, the frozen coating remained.

In July, when conditions were favorable, the Canadian Coast Guard did an ROV survey of the bow section, which was resting upside down in about 360 feet of water. Since the loading hatches were unreachable and the watertight bulkhead at the fractured end was pretty well intact, the ROV's operators had to content themselves with examining the outside. Close-up images of the fracture's edge revealed a number of small, clamshell-shaped fissures. Such fissures are typical of metal fatigue. Even though they were corroded from six months' submersion in salt water, they showed that the steel of the deck had been weakened for an unknown period previous to the accident.

The *Flare*'s EPIRB was never located. That no signal was received from it indicated it either did not break free or did not turn itself on as it was supposed to. This lack of a clear distress signal meant that the SAR teams took longer to locate the wreck, although it is not certain that they could have reached the site soon enough to save any of the crew even if the EPIRB had functioned properly.

* * *

For these three ships, all overwhelmed by fierce conditions at sea, it was small clues—such as the patches of snow and ice on the deck of a broken ship, the absence of a ventilator cap, and the nature of the ragged edge of a fractured hull plate—that provided vital information about the causes of the sinkings. The chapters that follow will explore further how deep-sea investigators put the pieces of maritime puzzles together.

3 The Ship that Toppled a Government: The Strange Case of the *Lucona*

Late on a quiet afternoon on January 23, 1977, the eleven-year-old Panamanian freighter *Lucona* was chugging uneventfully across the Indian Ocean, some 200 miles north of the Maldive Islands, when a violent explosion in her cargo hold ripped her hull apart. The *Lucona* was supposedly carrying a valuable cargo of uranium-ore processing machinery from the Italian port of Chioggia to Hong Kong—so valuable, in fact, that its owner, the Austrian businessman Udo Proksch, had insured it for the equivalent of $18 million. Within minutes, the costly assemblage went to the bottom along with the ship. The *Lucona* heeled slightly, righted herself, and then went down by the bow. The survivors testified later how they saw the smokestack, the wheelhouse, and the stern disappear one after the other. The *Lucona*'s massive propeller was still dutifully turning as the stern sank beneath the surface. Six of the twelve crew (including the captain's wife) survived, clinging to life rafts until a Turkish tanker sighted them and picked them up ten hours later.

Complications arose when Proksch filed his claim for the value of the cargo. In Austria's tangled political world, cozy affiliations between political parties and corporations made for interesting business dealings. Proksch was intimately involved with the governing Socialist Party and with its leaders. The insurance company, the *Wiener Bundesländer Versicherung*, or Viennese Provincial Insurance Company, was owned partly by the rival Austrian People's Party (the

Catholic-Conservative Party)[4] and partly by the Catholic Church itself as a business investment. It also had some private investors. The Provincial (as the Austrian media chummily called it) had been happy to accept Proksch's lucrative premium when the deal was signed in 1976. It was, after all, his house insurance company, and he maintained a running account there of over 70 million Austrian schillings.

In 1976, Proksch had taken out a policy for an upcoming industrial shipment: a uranium-ore processing mill that was to be delivered to an unnamed client in the Far East. Proksch gave the value of the uranium mill as 31,360,725 Swiss francs. The premium was commensurately large. When the claim was presented, the Provincial Insurance Company suspected fraud—Udo's reputation was rather shady—and balked at paying out such a huge sum. A Provincial Insurance subsidiary named Kasko paid the Dutch owners of the *Lucona* 3 million schillings for the loss of the ship. But the head office refused to pay Proksch for the alleged value of the cargo. Proksch brought suit, but the decision went against him.

The lawyer for the Provincial, Werner Masser, was one of the behind-the-scenes powers in the "Black," or Catholic-Conservative network. He was also the lawyer for a number of influential "independent" newspapers. Since Proksch's ties were to the Socialist Party, the ingredients for a major political quarrel were in place.

Udo Proksch was no ordinary Austrian. An immigrant from East Germany, he had managed to accumulate a large fortune by sharp (some said devious) business tactics and a loutish sort of charm. Among other business enterprises, Proksch owned Vienna's most illustrious *konditorei,* Demel's, whose luscious pastries were known throughout the gourmet world. (A *konditorei* is a sort of restaurant specializing in pastries and coffee, a very European institution.) For 200 years, Demel's had been the appointed purveyor of pastries to the Royal and Imperial Court of the Hapsburg Empire, until World War I put an end to that dynasty.

---

[4] Neither the Socialists nor the Catholic Party had a majority in Parliament, so for years they governed in an uneasy "Red-Black" coalition. At the time of the incident, the Socialists were the dominant partner.

Many of Demel's patrons were members of Vienna's wealthy elite. Above the public rooms was another Proksch enterprise, the private and very exclusive Club 45, frequented by the rich and powerful. Leading members of the ruling Socialist Party were said to belong to the club, and some suspected that Proksch also had links with the Soviet Union's secret police (the KGB) and Middle Eastern arms dealers. Thorough in his scheming, Proksch also wove a network of connections within Austrian military circles, which he exploited to the full. Foreign dignitaries were also among the guests at Club 45, including Imelda Marcos, wife of the dictator of the Philippines at the time, and Manuel Noriega, then dictator of Panama. Proksch and Imelda were friendly enough to dance together at the Vienna Opera Ball (Imelda kept a photograph of the happy occasion when she went into exile). The co-founder of Club 45 was a powerful Austrian politician named Leopold Gratz, who at various times served as foreign minister, president of the Austrian Parliament, and mayor of Vienna.

Physically unprepossessing, Proksch was a short, fat, grubby man with porcine features. His shirts were usually streaked with sweat. Yet he seems to have been a kind of tubby Svengali, seducing women right and left. He was married four times, once to a star of the Vienna City Theater Company and once to a great-granddaughter of the composer Richard Wagner. He was known to have made countesses pregnant, and his illegitimate children were allegedly legion. He also seduced sober businessmen into dubious deals.

Proksch's background might be called bizarre. He was born in 1934 in Rostock, Germany, to a family of Nazi loyalists. When he was a little boy, his father sent him to a "Nazi-factory." He hated the discipline and used to wet his bed, he said, to get sent home. "I was the smallest in the class," he told friends, "but the strongest." Still, he was at the bottom of the pecking order. But one day the great Heinrich Himmler himself, head of the dreaded SS and Gestapo, visited the school and transmitted greetings to Udo from his "brave" father. At once, the scorned little kid became one of the big boys. At this Nazi youth camp, Proksch learned that only a few people have anything to say—the members of a small heap of elite, as he put it.

Udo dropped out of school after the war and trained to be a

swineherd in Communist East Germany. Escaping to West Germany, he became a coal miner in the Ruhr. He also worked for a time as a corpse washer. Moving to Austria, he worked at a plastics factory that specialized in fancy eyeglasses. He acted in movies. He became an expert baker. He talked his way into public relations and set himself up as an industrial designer. With another German expatriate named Hans Peter Daimler, he built up a large business empire characterized by secretive and often fraudulent deals.

Daimler was as quiet and restrained as Proksch was noisy and flamboyant. He wore dark suits and gold-rimmed glasses and had an icy demeanor. At the trial, some of Proksch's associates said they feared Daimler and apprehensively pointed out how easy it would be for him to arrange a fatal automobile accident—"much easier than shooting," said one of them. Daimler was apparently the brains of the partnership. He knew enough about science and technology to sound convincing, and he was a master of creative bookkeeping. Proksch had a surplus of crazy and inventive ideas but was embarrassingly short on know-how.

For six years, the Provincial and Proksch brought civil lawsuits against each other with no result. But the Provincial hired a Swiss private detective named Dietmar Guggenbichler to dig up dirt on Proksch, and he found plenty. Guggenbichler was a man of pronounced right-wing views and therefore viscerally against anything connected with socialism. He was also indefatigable in pursuit of his quarry. Guggenbichler possessed a vain streak, too, for he made a videotape that he took with him when he traveled, in which he presented himself as an ice-cold snooper who was lightning quick on the draw. The Vienna newspapers liked to call him Dirty Dietmar, after the American movie hero Dirty Harry.

Within a few weeks, Guggenbichler had compiled an extensive file on Udo Proksch, including evidence that the allegedly valuable cargo on the *Lucona* was actually the wreckage of a conveyor belt from a closed-down coal mine. His file also claimed that Proksch had blown up the ship by means of a spark or timed explosion to collect the insurance. Both allegations turned out to be true.

"Dirty Dietmar" was thorough if not subtle. Once, shadowing the Austrian foreign minister (a close friend of Proksch's) on behalf of the insurance company, he stationed himself in the lobby of a Swiss hotel and photographed a "secret interview." His report to his client read: "19:15 hours: through the entrance stepped a man whom we identify as Mr. Udo Proksch. To be completely sure, we wanted to verify this with a telephone call. However, this became superfluous when two more persons entered, to whom the presumed Udo Proksch was introduced as Mr. Udo Proksch."

In 1983, Guggenbichler delivered his extensive collection of evidence to the district attorney of Salzburg, who brought charges of murder and serious fraud against Proksch, Daimler, and a third party. He was afraid that if he had gone to the authorities in Vienna the case would have gone nowhere, for Proksch's influential friends would have applied pressure on the courts to quash the proceedings. Nevertheless, the D.A. in Salzburg forwarded the case to the proper authorities in Vienna, where the investigation was sidetracked. Instead of being treated as a "preliminary investigation" with an independent investigative judge, it was handled under "preliminary inquiries," a category in which every step had to be approved by the attorney general. Udo's friends saw to it that the approval was not forthcoming.

Udo's alliance with the Socialist Party dated from 1971, when he bought Demel's (typically, he used friends as fronts in his business dealings; in the Demel's purchase, it was a countess). He announced, "Now the proles have taken over the rudder. I'll give them what they don't have: a place where they can dance, gorge themselves, and booze it up—but they'll dance to my tune." His Club 45 became the place where political deals were struck and where careers were made and unmade while its smiling host plied his friends with excellent coffee and pastries, food, and wines. It paid off.

Udo loved playing the role of court fool for Viennese high society and Socialist grandees. His pranks and follies were legendary. Claiming that Austria's cemeteries were overcrowded, he founded the Vertical Burial Society with the idea of saving space by burying people vertically in transparent plastic tubes instead of horizontally in coffins. In place of a costly tombstone, the head of the deceased person would protrude above ground. He invented a toothpaste tube that could be squeezed from both ends, calling it a marriage saver. He was known for shooting champagne glasses off serving trays at Vienna's better restaurants. All this, and more, was pardoned because he was so amusing. Some wondered, however, whether he was playing the court fool for the rich and powerful or manipulating them with these antics for his own amusement.

In 1984, the judge handling the investigation complained that his hands were tied. Every application for an official preliminary investigation was rejected by the minister of justice (a friend of Proksch's), who said in a TV interview that "the soup is too thin for that." In July of that year, the judge asserted his bold independence and had Proksch's and Daimler's premises searched. Authorities seized hundreds of pounds of documents. The judge's actions were blocked by Interior Secretary Blecha, who sent a Teletype message to the police to put away, immediately, all inquiries into the Proksch case. The judicial establishment complained loudly, and Blecha was forced to cancel his order the same day.

On February 1, 1985, the case took a new turn. The judge issued a warrant for Proksch's and Daimler's arrests on the grounds of the danger of flight and obstruction of justice. The reason: Proksch had told the court that he was going to Jakarta on business—but then had never made the trip, a sign of evasiveness. The devoted employees at Demel's marked the news of "Mr. Udo's" arrest by hoisting a black flag over the entrance.

The judge became a hero to the Viennese public, which was hungry for sensational news about the case. With four police officers in tow, he stormed off to Salzburg to deliver the decisive evidence. A scrap dealer there, named Günther Voglstätter, had testified several times that he had bought and scrapped the mining equipment Proksch had allegedly sunk on board the *Lucona*. The judge threatened him with arrest if he "did not finally tell the truth." The scrap dealer could tell no more than he already had.

Back in Vienna, the disappointed judge found that Foreign Minister Gratz had hastened to the aid of his friend Udo, to whom he sent an encouraging letter in jail. "Keep your head up," the powerful statesman urged. The letter reached the court records and, shortly afterward, the press. The judge commented sarcastically, "The lawyers can take a look at the documents any time. A photocopy costs six schillings."

Gratz called in the press and admitted that he had visited Proksch's storage shed at the Italian port of Chioggia (a look at a map will reveal that Chioggia, just south of Venice, is a very convenient port for shipping goods from Austria). There he had seen metal bars, piping, machine parts, and wooden crates but couldn't identify any of the equipment. The chief prosecutor read this tidbit in the morning paper and summoned Gratz for questioning. The foreign minister's testimony on the stand was evasive and uninformative.

Then a document that had been sought for eight years mysteriously surfaced. It purported to be the uranium mill's certificate of origin from Romania, delivered by special courier from the Austrian mission in Bucharest to the foreign ministry in Vienna and from there to the investigating judge. It looked like a crude forgery. The "evidence" from Romania surprised everyone, for Udo wanted to keep the origin of the goods he shipped secret. That was the way things were done in the international equipment business, he explained. As a German magazine put it, that business involves deals in the millions that whirl through a carousel of middlemen; briefcases full of banknotes, which are shifted from continent to continent as "useful payments" (translation: bribes); and much paper and few facts, an impenetrable wall of pretenses.

Proksch had set up a chain of shell companies through which the order for his uranium mill was placed by one of his confederates. It was as false as the rest of the plan, but it was convincing enough for the Provincial Insurance Company. After acquiring the antiquated coal conveyor belt, he stored it in a junkyard that he maintained in a small Austrian town named Piesting. There it joined other items that Udo had collected, including a worn-out grain mill and parts of jet engines scrapped by the Austrian Air Force. Using a loan that the Austrian government had given him to build a marmalade factory in this rural community, he used the money to erect a high fence around the property. He brought in Italian laborers to do this work—not a popular move in xenophobic Austria. When the fence was complete and hid his property from prying eyes, he had the Italians scrape the rust off the old machinery and rejuvenate it with a new coat of paint. When he was ready, he had it packed into containers and wooden crates, loaded onto a fleet of trucks, and shipped off to Chioggia.

There it was loaded carefully onto the *Lucona*, with the placement of each item personally directed by Proksch himself. The *Lucona*'s captain later testified that Proksch had been very careful to stow certain crates beside particular bulkheads. With Proksch was his good friend from the military, Major Johan Edelmaier, an explosives expert. Edelmaier had supplied Proksch with 880 pounds of "surplus" explosives left over from a training film that Proksch had made for the Austrian Army (filmmaking being another of Proksch's many ephemeral enterprises). It was later brought out that Edelmaier had furnished these explosives on the orders of another friend of Proksch's, Defense Minister Karl von Lütgendorf, or "Lü" to his intimates.

Proksch, a self-styled "weapons nut," loved the military. Thanks to his connections, he was allowed to ride as a passenger in combat aircraft and in tanks. Once, when drunk, he drove a tank down a civilian road and left it in the middle of a town. And he was never without his "Boomer," a 9-millimeter pistol that he kept tucked in his waistband.

In 1976, Proksch detailed Major Edelmaier to take care of loading the explosives onto the *Lucona*. Proksch even took him to the harbor in Chioggia to show him the ship. Then the cargo was insured for transportation but never checked by the insurance company. At the Italian border, it had to clear customs, but thanks to a corrupt official it again escaped inspection. Finally, it was loaded aboard the *Lucona* but apparently not weighed—it proved impossible afterward to establish its weight within 100 tons.

Proksch had another enemy besides the detective Guggenbichler: a rabid right-wing journalist named Hans Pretterebner. For years, Pretterebner had painstakingly collected every possible bit of discreditable information on Proksch, and at the end of 1987 he brought it all out in a carefully documented book titled *The* Lucona *Case*. Despite its dry, pedantic style and ponderous size—672 pages—it was an instant best-seller. The publisher could barely keep up with the demand. The book and its revelations shook Austria's political establishment to its foundations. The Socialists went out; the People's Party came in. Gratz and Blecha resigned in disgrace. Lü committed suicide. It became imperative for the government to move against Proksch.

A warrant was again issued for Proksch's arrest, but, hastily warned by his friends, he was able to mysteriously disappear just before the police reached him. Udo surfaced again in Manila under the protection of his friend and patron Imelda Marcos, whom he said was also his business partner ... and perhaps more. In Manila, Udo was treated by a faith healer for various illnesses: a slipped disc, arthritis, and acute alcoholism. He recovered sufficiently to grant interviews to Austrian reporters, and he made phone calls to friends and allies in Austria. As a German magazine commented, the Austrian police had to take a great deal of trouble *not* to find him.

When Interpol picked up his trail, Udo had to move again. He spent some time in Bremerton, Washington, where he acquired an American girlfriend of peculiar tastes. (Among his effects when he was finally caught were photos of her, naked, with her pet python crawling around her thighs.) Udo had plenty of American money with him for expenses. But even in the small, peaceful community of Bremerton, he could not escape Interpol, which sent out an international "wanted" on him. The FBI telephoned Bremerton's police chief to pass on the bulletin. The chief had to ask, "How do you spell 'Proksch?'" But Udo, under the assumed name of Peter Moss, slipped away again.

On the move, Udo passed through Hong Kong and then flew to Britain. There, at Heathrow Airport, he told the immigration inspector that he planned to spend a few days in Britain and then travel on to the United States. When he showed her his passport, which gave his name as Alfred Semrad,[5] the alert inspector saw that it bore signs of tampering. The British authorities put the document under an

[5] The real Alfred Semrad was an odd-job man for the owner of a tavern. He told police that his passport had been stolen, probably at a pop concert where he moonlighted as a security guard, but he hadn't noticed the loss until the day Proksch returned to Austria because he kept it in a folder full of papers "in my little box." At age forty-eight, Semrad was seven years younger than Proksch, but he bore a slight resemblance to the fugitive.

ultraviolet light and discovered that the entry and departure stamps were forged, one with a wildly discrepant date. The signature had been altered. The passport photograph was 5 millimeters too small, and the raised seal on it was inaccurate. Udo, who practiced forgery as a hobby, had been too casual about this one. (Later an Austrian police official sarcastically compared the forged seal to something a child would make with a carved potato stamp.)

Proksch was held in the guardroom at Heathrow for eight hours while the British police conferred long distance with their Austrian counterparts about what they should do about this person who was carrying a passport with irregularities and an enormous sum of undeclared money: $400,000 in U.S. currency plus smaller amounts of Swedish, Japanese, Filipino, and Turkish money. From Austria came the information that Alfred Semrad had a criminal record with nineteen counts, though none was recent—Udo had not chosen the right person to impersonate. But he had done a lot of traveling as "Alfred Semrad," because the passport bore legitimate stamps that showed he had been in France, Turkey, and Spain as well as in the Philippines. For some of this period, he had been accompanied by a longtime girlfriend, Alexandra Colloredo-Mannfeld, a pediatrician.

In addition to the suspiciously large amount of money, a search of Proksch's baggage revealed a collection of nude photos of women, some performing sex acts. There were also some photos of Hitler. Much more important were pictures of prominent Austrian politicians. When the airport police on the Austrian end looked through the collection, they exclaimed, "Jeez! There's Kreisky! [a former chancellor]. Look! There's your old boss!" Proksch asked the police to get rid of these "private photographs."

Proksch was allowed to make a phone call from Heathrow. He called someone named Miller, who advised him to take the next plane to Vienna. Proksch was eventually put on a British Airways flight. His passport was taken away and entrusted to the chief steward of the British flight, to be handed over to the Austrian authorities. According to one account, the steward refused to do the task, so the passport was returned to "Mr. Semrad," but he was escorted by guards onto the plane.

On the flight, a bizarre encounter took place. A Viennese art dealer and a friend of Proksch's, Evelyn Oswald, was returning from London and had nodded off when she was awakened by a hard poke. She looked up and saw a stranger with dark hair, a beard, and sunglasses. He shoved a boarding pass in front of her face. On the back, he had scrawled, "Hello love [the only part of the message written in English], the dogs are waiting for me. The purser has my passport. You must take my attaché case. Udo."

As the stranger walked back to his seat, she observed his short stature, his gait, his movements, his custom-made orthopedic shoes. There was no doubt—it was Udo. Proksch passed his friend another note asking her to pick up his carry-on attaché case when the plane landed while he distracted the steward, adding, "It's full of money and documents." Unable to decide between friendship and duty, she picked up his attaché case as instructed but left it on the floor of the shuttle bus from the plane to the main airport building. A helpful stranger called out, "Madam, you've left your bag!" She handed it over to a customs agent, who broke it open with a screwdriver after she told him it belonged to Udo Proksch.

Something strange must have happened along the way, for the Austrian police reported only $40,000 of the $400,000 that the British had discovered. The pornographic and political photos were also "lost." Proksch almost managed to get lost himself in the bargain. He disappeared into the transit hall and bought a ticket on the next flight to Nürnberg, safely across the border in Germany. The plane was due to leave at 5:05 P.M. The police caught up with Proksch at 5:02 P.M., but only because the bus from the waiting room to the plane was late. The airport police chief, who was a friend of Proksch's and had enjoyed a yacht cruise with him, did not recognize the heavily disguised fugitive. He asked, "But you aren't Mr. Semrad, are you?" "No, I am Udo Proksch," came the reply. Friendship apparently had its limits, for the police chief detained his old pal.

It was no wonder that the police chief did not recognize Proksch. In Manila, he had undergone plastic surgery: his prominent nose was shortened and narrowed, and fat deposits were removed from inside his cheeks, giving him a roguish, dimpled look. He grew a beard and mustache and got a wig to disguise his bald head. He also had his eyebrows thickened by a transplant and his eyelids darkened by tattooing.

The *Lucona* case became a political minefield. With Udo now in custody, a new judge was put in charge of the trial. His name was Hans-Christian Leiningen-Westerburg, and he had a reputation as an eccentric. For one thing, he rode a motorcycle to court instead of a sedate sedan. He was known to lecture defendants and break into impromptu speeches in the courtroom. He endeared himself to the press, however, by treating them to coffee and pastries in his chambers when trials got boring. They dubbed him "Rambo in the Robe."

Udo Proksch's trial took eighty-six days. Merely to work up various scenarios of a hypothetical explosion in the *Lucona*'s hold took a month. There were witnesses and counterwitnesses and changes in Proksch's legal team. At one point, one of the state prosecutors, irritated beyond restraint, snarled at one of Proksch's female lawyers, "Shut up with your idiotic cackling!"

Viennese society took the trial as the great entertainment of the season. The courtroom was usually crowded, even when the testimony was excruciatingly dull, for "Mr. Udo" was given to bizarre outbursts that gave the proceedings a carnival-like air. Describing himself as a "weapons nut," he pretended to fire an automatic rifle at one of the prosecutors, with accompanying sound effects. When asked what jet-engine fuel injectors were doing in his junkyard, he replied, "I collect hearses too." Proksch made contradictory statements and concocted colorful lies about his exploits as a mercenary soldier. He expressed his opinions of his ex-wives (he had four of them), declaring that he would rather sit at home face to face with a monkey. Now and then his defense lawyers had to scold him into proper courtroom behavior. He rambled on about trips he had taken and the short stature of many of his business associates, and he made a dramatic claim that he had had a "foretaste" of the next life.

At one point, Proksch launched into a convoluted tale about his connection with a right-wing paramilitary group that sought to take over Austria. At times, he grew bored and fidgeted in his seat; at other times, he stretched out and appeared to go to sleep. He fixed prosecutors with a hateful glare. Newspapers speculated that he was trying to convince the judge and jury that he was insane in order to evade a criminal sentence. One of his principal witnesses was caught in an outright and stupid lie. Others implicated him in fraud.

Proksch's defense team came up with imaginative suggestions to explain the *Lucona* disaster. One was that the ship had never sunk at all but had been captured by pirates and now, slightly disguised, was sailing the waters of the Far East under a different name. From the 1980s on, amid a resurgence in piracy, this would have been a possibility, but in 1977, when the *Lucona* sank, it was improbable. A more realistic defense was that the *Lucona* had struck a floating wreck beneath the surface—invisible but as deadly as a reef.

Outside the courtroom, rumors flourished. Some said the Russians had torpedoed or sabotaged the *Lucona* to keep the uranium mill out of American hands, while others claimed the Americans had destroyed the vessel to keep the Russians from getting this potentially dangerous equipment. Another possibility was that a bomb had been attached to the outside of the ship; perhaps one of Proksch's arms sales had gone bad, and the ship had been blown up to teach him a lesson. The anti-Semitic set claimed that the Israeli secret service was behind the sinking. None of these theories, as it turned out, was true, but the stories kept the Viennese public rapt.

A parade of witnesses gave testimony in the Vienna courtroom. A mechanical engineer stated that more than 80 percent of the essential parts of Proksch's alleged uranium-processing mill were missing. Proksch replied that it was indeed not a complete assemblage, only bait for bigger deals to come later. The bookkeeper at one of Proksch's firms told the court that all she did was sign papers without checking them—"Mr. Udo" himself apparently handled all the money transactions. She said she had only heard about the uranium mill from the newspapers. A carpenter said that he had built sixteen crates for the machinery but could not say what was put into them.

Proksch had previously claimed that the uranium-processing machinery came from Romania by forty trucks. He was forced to admit that he had used forged papers for the freight "because he had to have something to show the insurance company." But he could not produce any documents because of the strict secrecy of his deal, he said. A champion parachute jumper named Johan Huber, who had been Proksch's all-around gofer and worked at the junkyard in Piesting, testified that every day he would unbar the entrance to the yard, let the workers go in and out, make purchases, and blindly sign receipts for the alleged shipments of machinery, "because Mr. Udo asked me to."

Major Edelmaier, unhappy to have been dragged into the proceedings, testified that he had placed a "mere" 880 pounds of explosives at Proksch's disposal for making the training film but then said that he had actually handed over only 44 pounds, which Proksch had returned unused. As for the detonators found at the junkyard in Piesting, they were no proof of anything, said Edelmaier, for they lay by the hundreds on the ground at the military shooting ranges and could easily have been brought to Piesting.

Detailed references about blowing up the *Lucona* were found in Edelmaier's possession. He said he had prepared them on the advice of his attorney years after the explosion to test the accuracy of the accusations against Proksch (and himself). That was also the reason he stated for his interest in the workings of video recorders, whose timers could be programmed weeks in advance, making them convenient detonators. He denied ever having seen or been on the ship.

The survivors of the sinking were then called to the stand. Time had dealt harshly with them: three had died in the thirteen years since the sinking, and the captain and his wife were now divorced. The ex-couple and the surviving mate were all Dutch, which meant that a translator was needed. Each testified in turn. They all agreed that Proksch had definitely been in Chioggia, that he had supervised the loading of the cargo, and that he had made a point of its exact placement. He had insisted on positioning an especially large wooden crate beside each bulkhead. The captain and mate also described a pair of yellow-painted steel cylinders, each about 16 feet long and 10 feet thick, about whose placement the otherwise genial Proksch became testy and demanding.

The captain testified last. After he finished, the three witnesses, together with the jury, watched a simulation of the explosion that the prosecution had created using a ship model. Three variations were shown, with the explosion in the stern, in the bow, and amidships. Then came the grand finale: an explosion with the charges placed at the tops of the bulkheads, which caused the hatch covers to fly into the air—precisely what the survivors had experienced. After the last scenario, the captain's wife exclaimed, "That was it exactly!"

The captain, who had been snoozing in his cabin at the time of the explosion, described how he had only his underpants and wristwatch on when the blast occurred. The shock threw him out of bed and into the wheelhouse, which was filled with brown smoke. When he peered out at the railing, normally 30 feet above the water, he saw that it was even with the surface. He tried to escape, but his foot was caught. Though he managed to break loose, he injured his leg severely. Finally clear of the wheelhouse, the captain found himself in the water and then beside a life raft. But the raft was tied to the sinking ship by its painter. "I thought, 'I'm going down a second time if I don't cut this line,'" the captain recalled. "There was an emergency knife on it. I pull it out, and it is kaput. The ship was already below water level and was pulling the raft down. Then, fortunately, the rope broke." In the water, he lost his underpants and was wearing only his Rolex when the survivors were picked up by the Turkish tanker *Sapem I* ten hours later. He had to borrow underwear from one of the Turkish engine-room crew.

The three Netherlanders also told of Proksch's out-of-the-ordinary request for the advance calculations of the approximate positions of the ship at various speeds. Proksch obviously wanted to know where and when the ship would be at the right spot for sinking. He also demanded radio reports of the *Lucona*'s exact position each day. The survivors also called attention to Proksch's inexplicable directions to slow down the ship from the second day on, which, in their opinion, ensured that the programmed explosion would take place in deep water instead of by the shallow coast.

Next the court heard from the explosives experts, with their computer models and calculations of how much explosive it would take to sink a ship. They said 110 pounds, but that figure was based on arbitrary assumptions. It was possible to sink a ship with 110 pounds of explosive, but whether it really did happen that way on the *Lucona* the experts could say only if they examined the ship. A discussion of seventeen types of explosives and their characteristics followed. The expert witnesses all agreed that the kind of explosion that sank the *Lucona* required professional expertise. Only 150 people in the Austrian Army had such expertise, among them Major Edelmaier.

How was the explosion set off? A radio signal was as good as excluded, because it would have required an antenna on the ship, and the *Lucona* had none. Did one of the crew do it? Improbable. Could the ship's gyrocompass have been used as a timer? For that, the ship would have had to travel a particular distance in a specified time. Ruled out. Chemical and mechanical detonators were, for various reasons, excluded as "extremely improbable." That left only detonation by an electronic clock, but the only proof of this theory was recent testimony in a concurrent case against Edelmaier, in which a soldier claimed that the major was known to have taken an interest in the timers of video recorders. The experts dryly declared that such clocks did not exist at the time of the *Lucona*'s sinking.

The defense objected that none of the survivors had originally testified to having heard an explosion. Another expert said that the lack of auditory proof did not mean there had not been one; the noise could have been muffled by the weight of cargo on top of it, which would have been necessary to guide the force of the explosion.

The defense lawyers, obviously hoping that it would be impossible to find and photograph the wreck of the *Lucona* in the depths of the Indian Ocean, repeatedly pressed for a search expedition.

In the end, Judge Leiningen decreed that the sunken *Lucona* had to be found and photographed to decide the case. An American firm, Eastport (now a part of Oceaneering Inc.), was selected to do the job and appointed as the court's expert witness. After several months of preparations, Eastport shipped two ROVs and their ancillary equipment to Singapore in early December 1990. One of the ROVs, *Explorer 6000*, had a depth capability of 20,000 feet; the other, *Magellan 725*, could handle depths down to 25,000 feet. Once in Singapore, they were installed aboard the *Valiant Service*, an aging 165-foot supply vessel normally used for offshore oil operations. Judge Leiningen, who flew out to join the expedition in January, remarked that Proksch's prison cell was larger and more comfortable than his quarters on the *Valiant Service.* Accompanying Judge

Leiningen were two explosives experts from the Austrian Army and a naval architect, come to view the evidence with their own eyes.

The search area was about 200 nautical miles northwest of Malé, the capital of the Maldive Islands, off the southwestern tip of India. No reporters accompanied the ship, but daily progress reports were sent to Austria by satellite link. For a while, the news was mainly about the rather unappetizing diet of tuna casserole and the dwindling supplies of German beer on board. According to the Austrian magazine *profil*, the judge had counted on combining his investigative journey with swimming and snorkeling and perhaps an outing to picturesque Malé. He had to shelve those plans, for this trip was strictly business; the Austrian authorities in Singapore eliminated all possibilities of combining pleasure with work by requiring a strict accounting of all expenditures of money and time.

Eastport had spent a good deal of time plotting a preliminary search area based on calculations provided by the captain of the ship that had rescued the *Lucona* survivors. The captain of the Turkish tanker had been awakened at 1 A.M. with news of six people floating by in a life raft. He estimated the wind, a gentle breeze, as no more than 0.5 to 1.5 miles an hour. Factoring in the current, which in that part of the Indian Ocean was pretty steady, the ten hours the survivors told him had elapsed since the explosion, and the position of a radio beacon on the tiny island of Minicoy, the expedition leader did a backward calculation to yield the approximate position of the sinking. The captain's estimate gave Eastport a starting point.

The search vessel arrived at its destination, 8°50' N, 70°30' E, on January 23, 1991, the fourteenth anniversary of the loss of the *Lucona*. The ocean floor at that location is very volcanic and full of submerged peaks, so Eastport's first task was to make a preliminary chart of the area. This was accomplished with a precision deep-ocean Fathometer linked to global positioning satellite (GPS) via the company's integrated navigation system. The chart showed that the search area was relatively flat but with large changes in relief. Undersea mountains bordered the area from its northeastern to its southwestern ends. Mountains are bad news for sea-floor searches, for even an object as large as a ship can easily be hidden behind an outcrop of rock or in an undersea canyon.

The next day, the actual search began. The crew carefully lowered *Explorer 6000,* with its 20,000-foot depth capability, into the Indian Ocean. Towed at the end of a 33,000-foot cable, it was "flown" 350 feet above the bottom in order to give its sonars as broad a picture as possible. A computer translated what the sonar "saw" into visual images, which were displayed on a monitor on board the mother ship. At the same time, the sonar signals were recorded on two backup disks as a fail-safe procedure.

The judge was fascinated by the technology of undersea search operations. He was impressed by the idea that a computer in the towfish could sort out the sonar echoes and transmit them instantly to the big computer on board the mother ship, where they would be stored on disk for repeated future reference. The sensitivity of the sonar ensemble, which could distinguish objects like hatch covers from other parts of the wreckage, was awesome. He spent hours watching the sonar images on the screen and peering through the eyes of the ROV's video camera. He listened, entranced, as Eastport's equipment pilot instructed him in the subtleties of maneuvering a towed sonar sled. If it is towed too slowly, it sinks; if towed too fast, it behaves like a kite in a stiff autumn wind and goes aquaplaning up toward the surface. Handling the cable was also a ticklish matter, he learned. It does not just tow the sonar sled but also contains all the control cables and glass fiber optics. And, although the cable contains so many vulnerable components, it is also the most heavily stressed part of the outfit.

After six days of scanning the sea floor with sonar, during which almost 430 square miles of bottom were covered, *Explorer 6000* located a field of wreckage. Pieces of the hull corresponded roughly with the size of the *Lucona.* The depth was approximately 14,800 feet, far deeper than any diver or most manned submersibles could go. At this depth, the pressure of the water is 6,437 pounds per square inch, or almost 3.5 tons.

ROVs have opened up vast areas of the sea that were formerly unreachable, and the rugged *Magellan 725* went into the water on February 5. (Presumably, the searchers had to wait for favorable weather before they could launch it.) The *Magellan,* able to operate at 25,000 feet of depth where the pressure is nearly 11,000 pounds per square inch, was rigged for this dive with two manipulators, sonar, color and SIT zoom for close-up views, and black-and-white cameras. An optical fiber system transmitted the data from the sonar and cameras to the computer on the ship.

As *Magellan 725* flew slowly over a field of light debris, it encountered the stern of the wreck. It was sitting almost upright, heeling over slightly to the right. The force of its fall to the sea floor had buried it to the main deck level in the soft sediment. As one observer wrote, waves of mud radiated out from the mutilated ship as though it were sailing across the sea floor. The mud also hid the name of the ship, so its identity could not at first be established.

Excitement mounted as the data from the ROV began to come in. Would the images show that the hull was bent inward, indicating a torpedo or an external bomb, or would the hull plating be bent outward, indicating an explosion inside? Was the ship even the *Lucona*, or was it some other wreck?

The images, gathered over eight days on site, were clear and conclusive. The force of the explosion had almost destroyed the forward section of the ship. The heavy steel plates of the hull lay in shreds. The forecastle and chain locker had been blown nearly 650 feet away from the rest of the wreckage—the length of two football fields and more. The damage was found to be most severe in the area where Proksch had insisted his cargo be stowed.

Unfortunately for Proksch, his cargo containers *and their contents* survived the blast. Videotape and still photos clearly showed his cargo code number, XB 19, and the name of his Swiss front company, Zapata. The contents turned out to be the old coal-mining machinery and scrap iron, as the prosecution had maintained, plus a worn-out flour mill.[6]

The explosives experts also took great interest in the front surface of the nearly intact stern. The outward-bent deformation of the metal and its ragged edges showed that the hull had been sheared off where the midship cargo crane was located, along the lines

[6] According to some reports, Proksch's cargo also included a machine for making plastic pipes and a laboratory model of a device for making electronic components.

where the hull plates had been welded together. This damage could only have been accomplished by a massive internal explosive charge. Udo Proksch's friend Major Edelmaier had done his work well.[7] Yet one of the things that impressed Judge Leiningen most was the view of a carton of cigarettes in the wreckage, intact despite the massive damage caused by the explosion and fourteen years on the sea floor.

Judge Leiningen returned to Vienna with a thick bundle of still and video photographs. The jury had no difficulty convicting the well-connected Mr. Proksch of insurance fraud and the murder of six people. He was sentenced to twenty years in prison and also was forced to repay the costs of the undersea search and documentation. When the sentence was read, the outraged Proksch shouted, "Heil, Hitler!" implying that he had been convicted by a court of Nazis. A year later, after much public outrage, his sentence was changed to life imprisonment.

In 2001, Udo Proksch died in a prison hospital after a heart operation at the age of sixty-seven. Unbelievably, he still had admirers who posted eulogies to him on the Internet.

[7] In an unexpected outcome of the trial, Major Edelmaier was cleared of the charges against him, perhaps because he had acted on the orders of his superior, Defense Minister Lütgendorf. But, significantly, he left the courtroom as *former* Major Edelmaier.

4 Not-So-Benign Neglect: The *Marine Electric* and the *Rema*

The *Marine Electric* was not a safe ship. Hastily built as one of the "Liberty Ships" for the U.S. merchant fleet during World War II, at thirty-eight years old she was showing her age. There were holes in the hatch covers, a hole in the hull, and large cracks in the deck. Nevertheless, seafarers sought to ship out on her. One reason may have been that she had a reputation for serving good food. But the main reason was that her regular route never took her much more than 30 miles from shore. "If we get in trouble, the Coast Guard will come out and rescue us" was the word among the crew. Also, the voyages were short, so the crew could go home to their spouses, companions, and families every few days rather than every few weeks, as was the case for crews on trans-Atlantic voyages. And the owners of the *Marine Electric*, Marine Transport Lines, never pressured the crew to overload the decrepit carrier. Many of the thirty-four men on board the ship for the *Marine Electric*'s last run knew each other from previous voyages, and there seems to have been a good spirit of camaraderie among them.

The celebrated Liberty Ships were built to carry troops and supplies to the theaters of war overseas in World War II. They were built hastily because German submarines and planes (and those of Japan in the Pacific) took a horrendous toll on these unarmed ships. Their lifetimes were not expected to be long. But after the war, the U.S. government found itself with a large number of these ships, now surplus, on its hands, and it sold them to private entrepreneurs. The *Marine Electric*, having escaped German torpedoes on the Atlantic run, was one of these ships.

She was built in 1944 as an oil tanker of the T-2 class (the class is significant because many ships of the T-2 class sank at sea when their embrittled hulls gave way). Her original name was *Musgrove Mills*. In 1947, with the war two years past, she was sold by the government to the Gulf Oil Company and rechristened *Gulf Mills*. She carried oil for another fourteen years until a commercial shipping line, Marine Transport Lines (MTL), bought her in 1961. MTL had big plans for the middle-aged tanker, involving radical surgery. Her midsection was cut out in a Boston shipyard, and a new, German-built midsection 387 feet long was welded in, bringing her total length to 605 feet. Along with the new midsection, she acquired a new name, *Marine Electric*, and a new career. She became a coal carrier and spent most of the next twenty-two years shuttling between the coal-loading piers at Norfolk, Virginia, and an electric power plant at Somerset, Massachusetts. It was a short run—only 422 nautical miles after she cleared the headlands of Chesapeake Bay, thirty-six hours each way.

The Liberty Ships had certain inherent weaknesses. They were welded rather than riveted together. Welding, if properly done, produces a watertight seam but also weakens the metal. Riveting, on the other hand, preserves metal's strength. Furthermore, if a riveted hull plate develops a crack, the crack stops at the edge of the plate and does not spread to the next one. This is not true of welded hulls. Welding, however, is cheaper and quicker than riveting, and time was of the essence in delivering war supplies overseas to fight the Nazi war machine. Quick and cheap were apparently the watchwords.

It is also reported that many of the Liberty Ships were built of "tired iron"—steel whose proportions of carbon and alloy elements were not designed to withstand the chill of seawater or the repeated bending and flexing of a ship's hull as it passes over waves. Iron and steel (which is essentially just doctored iron) gain their strength by being forged, which aligns the fine crystal structure of the metal in an orderly pattern. Stress eventually transforms the small crystals into large crystals, which have much less cohesive power. (To test the validity of this assertion, take a wire or a strip of metal and bend it back and forth repeatedly. Sooner or later, it will become brittle and break, sometimes distressingly soon.) Welding, too, disrupts the crystalline structure of the metal. The iron of the *Marine Electric* was very tired after thirty-eight years at sea, and her crew knew it. Some of them had a standing joke: "Do you think they'll be cutting her up for razor blades [scrapping her] after this trip?"

"Can't make razor blades out of rust" was the stock reply.

Even for short coastal voyages, the crew always embarked with a sense of trepidation. Occasionally, MTL would get a government contract for the *Marine Electric* to carry surplus grain to Israel. When these long, deep-ocean trips came up, some of the crew would arrange to take their vacations.

This was the situation when the *Marine Electric* lay at the coal-loading dock near Norfolk, Virginia, on a cold, rainy Thursday night—February 10, 1983, to be precise. First Mate Robert Cusick was supervising the loading, taking care that the loading chutes spread the coal evenly to avoid shifting and to ensure an even balance. Not only did the old collier have to be level from side to side, but the bow and stern also had to be level with each other. Cusick was an expert; he had been a mate since 1943 and a chief mate since 1949, and he had served on colliers for twenty-eight years, six of them on the *Marine Electric*. The coal—24,800 tons of it—was destined for an electric power plant near Somerset, Massachusetts, near New Bedford. Because the boilers it was meant to feed were designed to have the fuel blown in, the coal was ground fine, into pieces that ranged in size from large pebbles to powder.

The captain for this voyage, Phillip Corl, was standing in for the regular captain, James Farnham, who was on vacation. Corl had flown out from his home in Las Vegas, Nevada, for the voyage. Under him were First Mate Cusick; the second mate, Clayton Babineau; and two third mates, Eugene Kelly and Richard Roberts. At the age of fifty-nine, Cusick himself held a captain's license but had several times declined the opportunity to command a vessel. Partly, he did not relish the isolation from the crew that a captain must keep if he is to maintain discipline; partly, he knew that as a captain he would have to take trans-Atlantic voyages, and he did not trust the old rust buckets that MTL used for this purpose.

Ironically, MTL was known as one of the best shipping lines in the United States. It ran an honest operation, and it had a fleet of sturdy modern tankers and bulk carriers. It also was saddled with six decrepit ex–Liberty Ships, and economics compelled the line to get the last possible mile out of them before they went to the scrap yard. MTL was owned by a huge conglomerate, GATX, which wanted to have its MTL division turn a profit. Thus, there may have been pressure from top management to cut costs by skimping on maintenance. These details would come out later in the Coast Guard's investigation of the *Marine Electric* tragedy.

There was further irony in the *Marine Electric*'s being scheduled for dry dock after the February 1983 voyage. MTL's fleet manager had requested a postponement from the Coast Guard until April. Business-friendly, the Coast Guard had acceded.

It was a bad mistake. The ship's hatch covers were so badly rusted that they were full of pinhole cracks through which the crew could see daylight. There were larger cracks, too, some as big as 2 feet long and 3 inches wide. The crew tried to patch the cracks with duct tape and an epoxy cement called Red Hand, but new cracks and holes were constantly appearing. One day Third Mate Kelly discovered a 3-inch hole in the deck itself. Out of curiosity, he shined a flashlight into it. The hole went all the way through the heavy steel plate of the deck. Kelly himself refused to walk on the hatches for fear of falling through, though others in the crew took their chances.[8] None of these defects would have been fatal in itself, but all together they added up to danger.

---

[8] The hatch covers were 37 feet 10 3/8 inches wide, so that they covered half the deck from side to side. They were divided into six panels 6 to 7 feet long, loosely connected by chains, and ran on tracks. To open these MacGregor Single-Pull Hatches, a chain or cable was made fast to the forward panel, and winches and pulleys pulled the whole assemblage back. As the hatch panels reached the end of the track, they stacked up automatically, like dominoes. The process was reversed to close the hatches. Rubber gaskets around the edges of the panels, when in good repair, kept them watertight.

The crew and mates reported the damage to the chief engineer, who in turn reported it to management;[9] this was the way things were done in the merchant marine chain of command. To go outside the established channels meant risking your job—and, although a senior officer could earn $100,000 or more a year, none was independently wealthy. In the event, the company did nothing to repair the hatches.

Then there was the hole just beneath the bow, which Kelly had discovered while the ballast tanks were being filled after the ship had been unloaded at Somerset at the end of the previous voyage. Kelly had spotted water pouring out of the hole and called Cusick. The men concluded that the hole had probably been made by a bulldozer as it was being hoisted aboard the ship by a crane. It was to be lowered into the holds to push the coal into piles for easier unloading. Swinging on a cable, the corner of the dozer's blade would have banged into the hull and penetrated it. The hole was only 3 inches across and located 5 feet below the deck. This would have placed it well above the waterline in calm weather. But it would have to be repaired.

Cusick applied a temporary patch made of the bottom of a coffee can, which he cemented on with epoxy. He backed it with a box filled with cement, braced it with a timber, and reported the incident to the master of the vessel. The temporary patch was still in place when the *Marine Electric* embarked on her last voyage. The crew, in desperation, even outlined the worst cracks and holes with chalk to call them to the attention of the Coast Guard and classification society inspectors. The inspectors studiously ignored them.

The sea was running high when the *Marine Electric* cleared Chesapeake Bay and entered the open sea. A gale-force wind was blowing, and the air was thick with driving snow. But the ship had been out in worse weather before and survived. The waves soon increased to 25 feet. The *Marine Electric* plowed ahead at low speed,

[9] Normally, the permanent master of the vessel would have the ultimate responsibility for requesting repairs and making sure that they are carried out. However, the master of the *Marine Electric*, Captain James Farnham, testified repeatedly at the Coast Guard hearing that he had no authority over repairs and consequently did not concern himself with them.

barely making headway. Attempting to make speed under such conditions would have been suicidal, and Corl was a competent and conscientious captain.

When the *Marine Electric* had reached a position more or less off Assateague Island, it got a call from the Coast Guard. A fishing boat, the *Theodora*, was in trouble nearby, and the Coast Guard asked Corl to stand by and render assistance. The 65-foot *Theodora* was taking on water, and her pumps were either failing or just could not keep up. The Coast Guard was going to deliver a spare pump to the floundering fishing boat by helicopter, but that would take time; it would be desirable if the coal carrier could stay in the neighborhood to take the crew of the *Theodora* on board if necessary. The coal carrier's radio would also help the Coast Guard find the ships' location. Corl replied that his own vessel was having difficulties, but he agreed to the request, for it is the duty of sailors to help other sailors who are in danger.

Corl, at the helm, managed to execute a U-turn in the house-high waves so skillfully that his ship was in the troughs for only two or three waves. With the wind at his back, he steamed south toward the *Theodora*. In fact, the *Marine Electric* had passed the *Theodora* at about 3:30 P.M. on Friday, February 11, and the fishing boat had seemed to be holding its own at that point. The *Marine Electric* was only a few miles away when the Coast Guard called. Blinding snow changed to a violent rain squall, but by 4:36 P.M. the *Marine Electric*'s radar picked up the *Theodora* only a little over a mile away. The Coast Guard asked Corl if he could stand by until midnight. He agreed reluctantly, saying, "I'm having problems out here myself in this ... weather."

At 6:22 P.M., the *Marine Electric*'s captain notified the Coast Guard, "I'm taking an awful beating out here. I'm going to be in trouble myself pretty soon." Two minutes later Corl told the Coast Guard, "I don't know how I can hold ... heave to on this course. I'm rolling, taking water, green water over—over my starboard side, all the way across my deck."[10] A few minutes later the Coast Guard gave

[10] *Green water* is a seafaring term for a solid wall of water running over the deck.

him permission to leave. The Coast Guard helicopter had managed to land the replacement pump on the heaving deck of the *Theodora*; a Coast Guard rescue boat had made its way out to the scene, and the fishing boat was once more proceeding toward shelter.

The area of the Continental Shelf over which the *Marine Electric* now found herself was shallow; in some places, the shoals were only 40 feet deep. With her load, the *Marine Electric* was drawing 34 feet of water. In the trough of a 25-foot wave, she could easily bottom out. Even though the bottom was soft and sandy rather than rocky, the force of 28,000 tons slamming down on it would cause severe damage to the ship. (Have you ever been slammed into the bottom by a big wave while at the beach? Multiply that impact by several thousand times, and you have an idea of what a ship's hull would have to withstand.)

Just what happened to the *Marine Electric* is unclear, and may never be known, but about 1:15 A.M. on February 12, Third Mate Roberts, at the helm, noticed that the vessel was responding sluggishly to the waves. She seemed to be down at the bow, which should have been riding up over the waves. The situation did not look favorable. Roberts woke Captain Corl and summoned him to the bridge.

Paul Dewey, an able-bodied seaman, had taken the helm at 11:50 P.M., a little before he was scheduled to go on watch. At 1:00 A.M., he was given a break and went below to change into dry, warmer clothes. When he returned to duty a little before 2:00 A.M., he noticed the master and the mate talking about the trim of the bow.

Corl and Roberts worriedly looked out along the deck. Waves were breaking over the bow, a danger signal: with the ship so bow-heavy, water had to be coming in. They tried to order the engine room to start pumping but found that the ship's telephone had failed. A seaman was rousted from his watch station and sent to notify the engine room and round up those officers who were not on watch. Corl awakened Cusick and asked him to come to the bridge. "I believe that we were [sic] in trouble," he said. "I think she's going—settling by the head." He added that it might be his imagination; with the seas running so heavy, he couldn't tell.

Cusick took a look for himself. The captain was not imagining things. He ran to get the chief engineer, Richard Powers, who peered forward through the spray and dark. The three men conferred briefly and concluded that their situation was desperate enough to call the Coast Guard for help. Corl informed the Coast Guard that he was approximately 30 miles from the entrance to Delaware Bay and would try to make it there, but he was taking on water forward and going down by the head: "We are positively in bad shape. Positively in bad shape, we need someone to come out and give us some assistance."

The crew shined flashlights toward the bow to see what was happening; wave spray blotted out their beams. Powers ran down and fetched his big tankerman's lamp, which was powerful enough to reach the bow. Powers handed it to the man on lookout. By its beam, Kelly could see the "doghouse" on the bow, a small, white structure that gave access to one of the holds. It kept disappearing and reappearing above the waves. Kelly could not see the hatches but thought that the entire bow was under 6 or 7 feet of water. Powers thought that the No. 1 hatch (the one closest to the bow) had been staved in, though he could not tell because it was covered by water. When storm waves stave in the bow hatch of a boat, there is no hope left: the hold floods; the bow goes down; the other hatches are staved in one after another; and the boat sinks.

At 2:54 A.M., Corl radioed the Coast Guard to tell them, "I'm having my crew muster at the lifeboats." Shortly afterward he informed them that he would try to head for the entrance to Delaware Bay.

About 3:30 A.M., Corl told Cusick to prepare the lifeboats for launching. Merchant seafarers have little use for lifeboats. They are easy to launch in calm weather, but when the ship is rolling and pitching in a storm they become problematic, swinging wildly on their hoisting cables and crashing into the ship's side. Once in the water, they obey a rhythm all their own and not in synchrony with the endangered ship's, which makes it hazardous to try to board them. And, if the ship is listing toward one side, the boats on the high side get caught on the ship's hull, while those on the low side dangle out of reach. Furthermore, if survivors are in the water, it is almost impossible for them to get into one without major assistance.

The wreckage of the USS *Maine* in Havana harbor.

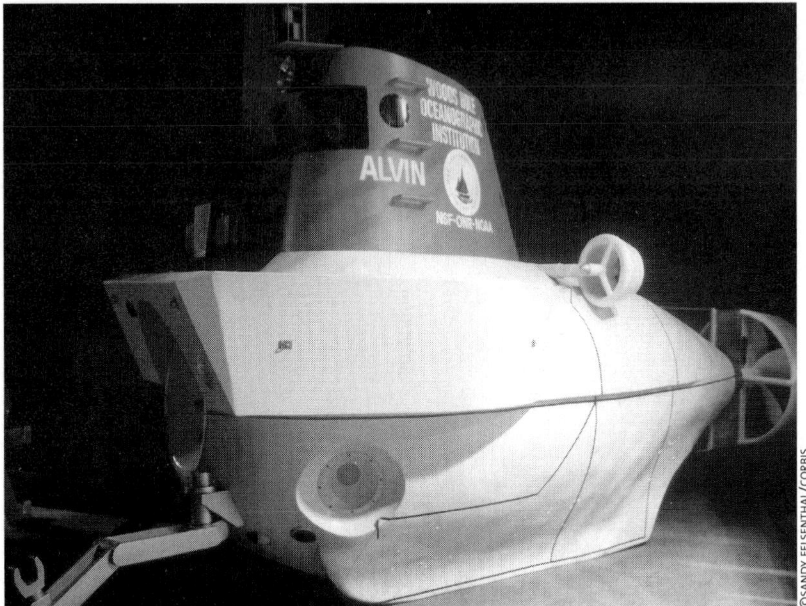

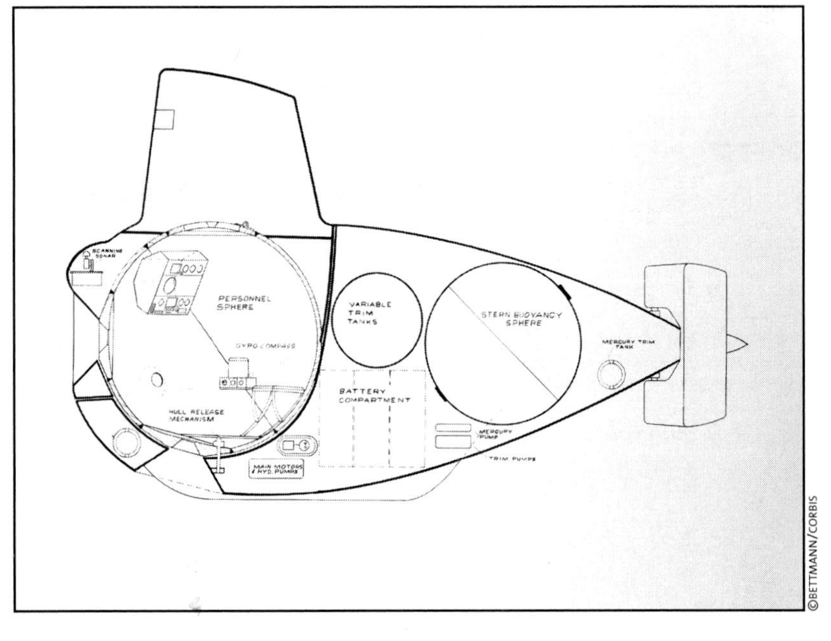

Two views of the famous submersible *Alvin*.

ASSOCIATED PRESS THE NEWS TRIBUNE

Two Newt atmospheric diving suits.

COURTESY OF THE TRANSPORTATION SAFETY BOARD OF CANADA, GOVERNMENT OF CANADA

The bow section of the sinking *Flare*.

Recovery of the visor of the *Estonia*.

An image of the wreck of the *Edmund Fitzgerald* is captured by *CURV III*.

Illustration of the sinking of the *Lusitania*.

The reassembled fragments of TWA Flight 800.

The black box from Swissair Flight 111.

The *Kursk* in dock in May 2000.

The lifting pontoon *Giant 4*.

ASSOCIATED PRESS/U.S. NAVY

The top of the bow section of the sunken USS *Scorpion*.

WOODS HOLE OCEANOGRAPHIC INSTITUTION

Abe, an AUV.

It is no wonder, then, that mariners sometimes refer to lifeboats as "death boats." They prefer the inflatable life rafts, although these have their own drawbacks.

Nevertheless, the crew lined up obediently on the deck, as they had been taught to do in lifeboat drills. Cusick had them take the covers off the lifeboats and fold them neatly so as to take up minimum space when stowed. As an extra precaution, Third Mate Kelly took the life rings from their locker and piled them on deck on his way to the lifeboat. He later said that he had never been trained to do it—bringing out the life rings was not a part of the lifeboat drill—he just did it without thinking. It was a fortunate reaction. Kelly also cut the lanyard of the ship's emergency positioning radio beacon (EPIRB) and stood it upright in its box—just in case.

On the bridge, Dewey was still at the helm. He could see and feel the *Marine Electric*'s bow going down gradually. A radio message came in from the Coast Guard: did the ship have survival suits? No was the answer—just life jackets. (A life jacket is not the ultimate in protection if your ship sinks. It keeps you afloat with your head above water, but it does not prevent you from swallowing water if a wave breaks over you, and it does not protect you against the cold.) The *Marine Electric*'s lifeboats were swung out over the water on their davits, but not yet lowered, as the ship was rolling too severely. The life rafts were brought down to the boat deck, still in their protective canisters.

Around 3:40 A.M., the *Marine Electric* again signaled that her time was drawing very near: she began to list to starboard. Another tense quarter of an hour passed, and her inclinometer was showing a worrisome list of 5 degrees. When the ship rolled in the waves, the list increased to 14 degrees, almost dunking the starboard lifeboats into the sea and making it difficult for the crew to keep their footing on the wet deck.

At 4:07 A.M., the list became more pronounced, and the *Marine Electric* tilted to 10 degrees. Captain Corl radioed the Coast Guard again, saying, "I think I'm going to lose my ship here ... We are taking a real bad list to starboard." He called the engine room on his walkie-talkie, ordering the crew down below to stop the engine and evacuate the engine room.

At 4:10 A.M., the radio operator came onto the bridge with the news that two merchant vessels were on their way to the scene. One was about an hour away; the other would reach the scene at about 8:00 A.M. As Kelly later testified, "it was kind of strange, because on the bridge everybody kind of let out a sigh of despair. They knew we didn't have that much longer to stay afloat."

Ordered by the captain to give up his attempts to steer, Dewey jammed the rudder hard to port and came down to the boat deck. Meanwhile, Corl sent out his last message: "We are abandoning ship right now! We are abandoning ship right now!" The Coast Guard logged this communication at 4:14 A.M. It had taken only seven minutes for Corl to reach his decision. Captains abandon their ships only with the greatest reluctance; they know that management does not like this course of action and does its best to penalize them if they take it. But sometimes making this extreme decision is necessary.

Kelly blew the ship's whistle, the signal to abandon ship, and began heaving life rings into the dark water. If the crew were lucky, the life rings would not drift away before they could reach them. Suddenly, the ship turned on its starboard side with a horrid sucking noise, flinging the surprised men into the water.

Cusick found himself trapped underneath the deckhouse, now lying on its side. As he later described the surrealistic scene for the Coast Guard's Marine Board of Investigation, "it was like the sound of the last water draining out of a bathtub, amplified a billion times, and I was clawing and swimming up. And I had been standing outside the 12–4 engineer's room,[11] and the lights were still on. I looked right in the porthole and swam by it. I kept swimming, swimming till I reached the railing. I turned it, and I shot up. I had on a pair of padded underpants, the quilted type, polyester in them. I believe they had a lot to do with my coming through.

"I come up, broached the surface, took a deep breath, and not far from me I could see the smokestack. It seemed to be just a little bit above the horizontal. I started swimming out."

[11] The 12–4 engineer had the watch from 12 midnight to 4 A.M. and again from 12 noon to 4 P.M.

Buoyed by his life jacket and padded underpants, Cusick swam for what seemed to him like half an hour until he found an oar to cling to. Each time he rose on a wave he could see the strobe lights of his shipmates' life jackets blinking in the darkness, and he heard their cries and groans, but he could not reach them. Finally, he dimly made out the shape of a lifeboat on the crest of another wave. Torn loose from the ship, it was drifting. Cusick labored in the deathly cold water for another half hour before he managed to reach it. It was swamped, which gave him a chance to get in. But in the heaving sea, he had to wait for his chance. Holding onto the gunwale with both hands—he had abandoned his oar—he kicked off his boots and waited. His chance came when a wave carried him and the boat up together. Then the boat started down while Cusick was still rising. He gave a heave and let the wave topple him in. It was not much of an improvement. Sitting on the gunwale with his buttocks only a few inches above the water, he felt the freezing cold of the air. A wave nearly washed him overboard, so he flopped back into the nearly filled lifeboat and did his best to keep warm by thrashing about. He yelled to attract attention. But there was no one to hear him. While he was in the lifeboat, Cusick compared its length, 24 feet, to the height of the waves. He estimated that some of them were 26 feet from trough to crest.

When daylight finally came, Cusick could see rescue at hand. A big Norwegian tanker, the *Berganger*,[12] was pulling alongside his lifeboat. The Norwegian crew dropped a Jacob's ladder over the side, and a couple of them cautiously made their way down. They tried to reach out to Cusick, but the waves were too high. The Norwegian captain, realizing that the waves might smash the lifeboat and its half-dead occupant into the side of his ship, backed away. Cusick later told an interviewer that he realized this was the right decision. Rather than deserting him, the Norwegians were trying to spare him. Suddenly, he heard the noise of a helicopter overhead. The Coast Guard had finally found him. The copter lowered a rescue basket,

[12] This is the way the name is spelled in the Coast Guard report. The National Transportation Safety Board (NTSB) renders it as *Bananger*, and the *Philadelphia Inquirer* spells it *Barranger*.

and Cusick, exhausted and nearly paralyzed by the cold, somehow managed to tumble in.

Inside the helicopter were his shipmates Kelly and Dewey. These three men were the only survivors. Cusick was in desperate shape when he was finally pulled in. He was covered so thickly in heavy oil that at first his shipmates could not recognize him. Their wits dulled by the cold, they thought he must be one of the African-American members of the crew. "Was there a black man with a beard on board?" one asked the other. Then they could not tell whether he was dead or alive. As the Coast Guard tried to resuscitate him, they repeatedly asked him what month it was. (This was a technique for determining if he was alive and conscious.) Finally, Cusick, coughing and vomiting up a stomachful of seawater and fuel oil, managed to reply that it was February.

Dewey and Kelly had endured harrowing experiences of their own. Dewey had also been trapped underneath the ship when it turned over, but he had managed to swim clear. Swimming away from the ship on his back, he could hear men around him crying for help. He felt a line as he swam, reached for it, and found it was attached to the canister of a life raft, which had failed to open on contact with the water. Dewey braced his feet against the canister and pulled on the line as hard as he could. The canister popped open, and the raft self-inflated, as it was meant to do. But in so doing, it blew Dewey off. As he swam back toward the raft, he could make out three other crew members close by. The men struggled for fifteen to twenty minutes to climb over the 3-foot-high sides of the life raft, hampered by their heavy clothing and the numbing cold (39 degrees in the water, 29 degrees in the air). Dewey finally made it into the raft, but another seaman whose name he did not know was too weakened to make it on his own. Dewey tried in vain to pull him in as waves washed over them, but the man was beyond help.

Dewey yelled to the other two seamen to hang on to the lifeline around the side of the raft. Then Clayton Babineau, the second mate, swam up out of the darkness. Dewey could not pull him into the raft, either, despite Babineau's efforts to help. Babineau told Dewey to put the raft's rope ladder over the side; if Dewey could help him get up the ladder, then he could help Dewey pull other men in. But the raft lacked a ladder. A cargo net, on the far side of the raft, would have to substitute. Dewey urged the men to work their way around the raft, hand over hand, to the net. Babineau, first to get there, tried to climb the net, but his hands no longer functioned, and he could not get a grip. His weight had pulled the net flat against the top of the raft's sidewall, and there was not room for him to get a grip in any case. Dewey resourcefully gathered the net into a bunch so that Babineau could grab it, and he even placed Babineau's hands on it. The exhausted mate still could not manage to hoist himself onto the raft. Dewey told him to get a foothold in the meshes of the net; Babineau cried that he couldn't. Then, no longer thinking clearly, Babineau put his feet over the side of the raft. Dewey, yanking on his ankles, managed to get Babineau on board as far as his knees, but that put his head under the water. The attempt was abandoned.

Dewey searched frantically around the raft for something that he could use to help. When he looked back, Babineau was gone—the second mate had lost his tenuous grip and floated away out of sight. That left two other seamen in need of help. One was conscious enough to try to get into the raft, but again Dewey did not have the strength to help him. The two were too exhausted to do more than just hang on. Dewey remembered that they kept crying, "Help me! Help me!" They, too, lost their grips and drifted off, leaving Dewey alone in the raft.

He heard the noise of a helicopter and shined his flashlight toward it. The helicopter passed him, made a circle, and came back. It lowered a basket, and he managed to climb in. Pulled up to safety, he managed to shout that there was no one else in the raft. The unspoken message was that the Coast Guard should not waste their time looking for other rescuees in that spot. But when he looked down, Dewey saw a navy swimmer in a wet suit, fins, and snorkel in the water.

When the helicopter had reached the scene of the accident, its crew had felt elated. The water was full of men in life jackets, apparently waving their arms to attract attention. The helicopter would save them! In reality, they had succumbed to the cold and to exhaustion. Dead, they floated in their buoyant jackets, their arms flopping up and down with the motion of the waves. The navy swimmer found only one more living person from the crew of the *Marine Electric*: the third mate, Eugene Kelly.

In some ways, Kelly may have had the most frightening experience of all. He had prepared in good time for the emergency, putting on heavy woolen clothing (wool is a good thermal insulator even when wet) and packing his valuables in a knapsack. Into the knapsack went Kelly's two cameras, binoculars, a knife, and his transistor radio. His wallet, eyeglasses, and car keys went into the pockets of his clothing. When he was shaken awake by one of the crew at 3:00 A.M. and told to report to the bridge with his life jacket on, he washed his face and brushed his teeth and left the knapsack behind on his bunk. But when the moment came to abandon ship, he had no time to retrieve it.

As he was frantically making his way to the lifeboat, his walkie-talkie emitted the voice of the duty engineer down below. The engineer asked if the officers wanted the engine-room pumps tied down. "Get the hell out of there!" Kelly yelled back. "We're going down!" Kelly fell going down the stairs but recovered and made his way onto the deck. As the ship tilted more and more, he slid into the water. It happened that he was directly under the ship's smokestack as it descended over him. Disoriented, he stared at the stack as it loomed larger and larger. At the very last moment, the hand of a shipmate whom he could not see yanked him out of the way by the back of his life vest. Before Kelly could see who had rescued him, the man was gone.

Like Cusick and Dewey, Kelly thought he swam for about half an hour in the inky blackness. Eventually, he came upon a life ring floating on the water, with five other men clinging to it for dear life. Three of them he knew: the other third mate, the chief engineer, and the radio operator. The other two were new to the ship, and he had not learned their last names yet. Kelly had the men sound off by number so that they could ascertain in the darkness how many of them there were on that single life ring.

As they clung to the ring, they heard a loud explosion; Chief Engineer Powers said it must be the boilers exploding. For some time, the men talked encouragingly to each other, but one by one they began to slip away, complaining of the cold. The chief engineer had stopped shining his powerful flashlight into the air. Kelly thought that he might have lost it and whacked the man on the back of his life jacket to rouse him. There was no response. The chief had died without a sound. His flashlight began to float away, but Kelly managed to get hold of it and used it to signal their position. The last to go was the radio operator, just as the Coast Guard helicopter arrived and lowered its rescue basket. When Kelly was hauled up, he vomited seawater and oil from the *Marine Electric*'s fuel tanks and sobbed uncontrollably. Captain Corl was gone without a trace—Kelly had last seen him climbing the rail of the ship while struggling with his life jacket. The engine-room crew, intent on doing their duty to the last second, had been trapped below decks. There were many fellow crew members to mourn.

The Coast Guard had been too late to save most of the *Marine Electric*'s crew. It was not for want of trying, but a number of things had delayed them. First there was confusion over which Coast Guard station had jurisdiction over the rescue: Ocean City, Maryland; Indian River, New Jersey; or Cape May, New Jersey? The Coast Guard believed the *Marine Electric* was close to Cape May, on the northern shore of Delaware Bay. But when the floundering ship radioed its last position, they realized that she was only 9 miles from where she had stopped to help the *Theodora*—the severe weather had not allowed her to progress farther. The ship was in Ocean City's bailiwick.

Ocean City had no helicopters on standby. Cape May had some old, short-range, lightweight helicopters, each of which could hold six people and whose best speed was 90 knots under good conditions—a speed certainly not possible when fighting through 50-knot winds. Elizabeth City, North Carolina, had modern helicopters that held fifteen souls; their twin engines could push them through the air at over 200 knots. But Elizabeth City was 100 miles away from the accident scene. A decision was finally made to send a rescue copter from Elizabeth City—but it took time to arrive at that conclusion.

Problem two was that, in the storm, the Coast Guard needed assistance to find the *Marine Electric*. Three Coast Guard cutters had been dispatched between 3:15 and 3:30 A.M., but in the raging seas it would take them hours to reach their goal. The only hope of rescuing the men from their sinking ship was helicopters. But in the storm, no helicopters could pinpoint the target ship without guidance from another vessel, and there was no other vessel nearby. The Coast Guard had to arrange for a fixed-wing aircraft to fly cover for them, which took time. Then they thought that the *Marine Electric* might possibly be saved if they flew heavy-duty pumps out to her to get the water out of her hold. The pumps had to be fetched and loaded onto the copter. At this point, Captain Corl's message that he was abandoning the ship came in. The entire rescue plan was altered immediately. The pumps were no longer needed to bail out the ship; instead, thirty-four crew members had to be rescued from lifeboats or from the water. The pumps, now only useless excess weight, had to be unloaded. It was 4:13 A.M. before the helicopter from Elizabeth City got into the air.

Another problem was that the Coast Guard had no rescue swimmers. They had trained and expert divers, but the two skills are different. The Coast Guard diver on station pleaded to be allowed to go out and help with the rescue, but his superior officer forbade it. A hasty call was put in to an off-duty navy swimmer, James McCann, who responded promptly. But he had to get from his house to the navy heliport. More time slipped away. When the navy copter got to the scene, the Coast Guard was already there, in a quandary. Not knowing that the men in the water were dead, the Coast Guard wanted to pull them out but was not equipped to do so—which was why they had called in the navy. But the navy helicopter pilot, peering down in the darkness, couldn't see anybody, and he did not want to risk his swimmer's life for nothing. The Coast Guard insisted that there were men in the water, and the navy pilot gave in. McCann was lowered partway to scout out the scene. Dangling 20 feet below his chopper and 20 feet above the sea, he could see nothing until the helicopter's navigation light picked out the reflective tape on the men's life jackets.

Entering the water came as a shock to this seasoned swimmer. Even through his wet suit, the cold struck him hard. For ten minutes, he could only gasp for breath and cough out water that he had accidentally swallowed. He made so little headway in the violently heaving waves that the Coast Guard copter had to ferry him over to the bodies in its rescue basket. Valiantly, he grabbed one man in a life jacket after another, lifting them into the basket. Even if they were dead, they deserved to have their bodies recovered. And he could not give up until he was sure that no survivor was left in the water. When he rescued Kelly, he could not tell whether the man was alive or not. At length, McCann's endurance gave out, and he had to be pulled in, though he protested that he must continue his search.

McCann and the three survivors were swiftly dispatched to the hospital, where they were cleaned off and carefully warmed up to avoid the complications that ensue when a victim of hypothermia is warmed too quickly.[13] McCann, young and in excellent condition, was back on his feet in a few hours; miraculously, the other three recovered quickly. It was noted that they were the fattest men on the ship; medical researchers concluded that their layers of body fat had insulated them from the worst effects of hypothermia. Cusick, aged fifty-nine, also owed his survival to a Thinsulate jacket that his wife had bought him and that he was wearing when the ship went down.

A word here about hypothermia: this condition is defined as the lowering of the body's core temperature (the heart and lungs) to about 95 degrees Fahrenheit or less. Impending hypothermia is when the core temperature starts to fall below 98.6 degrees Fahrenheit. At this stage, shivering sets in: it is the body's attempt to warm itself by generating heat from rapid muscle contractions. Complex tasks with the hands can become difficult.

Safety authorities define moderate hypothermia as beginning at 95 to 93 degrees Fahrenheit. At this point, there is intense shivering, and victims have trouble coordinating their movements. Motion becomes slow and clumsy, and the mind may become confused. When the body's core temperature drops below 93 degrees Fahrenheit, victims typically have trouble speaking, and thinking becomes sluggish. And below the range of 90 to 86 degrees Fahrenheit, severe hypothermia sets in. As the body's temperature continues to sink, victims progressively become very confused, may behave irrationally, and lose the ability to talk coherently. Eventually, they lose consciousness

[13] If a hypothermia victim is warmed up too quickly, blood drains away from the heart into the outer parts of the body. This deprives the heart itself of oxygen, which can send it into fibrillation or arrest.

and, if not treated in time, suffer cardiac and respiratory failure—in other words, they die.

The same safety authorities also have advice for victims of boating accidents: the most important thing to do is to conserve body heat, which is lost thirty times faster in water than in air. If you are alone, hug yourself around the chest and draw your knees up into a fetal position. If you are with other people, get into a tight huddle. Do not try to warm yourself by treading water or swimming—those actions will make you lose heat faster.

The Ohio Department of Natural Resources, in its guides to water safety, states that, in water between 40 and 32.5 degrees Fahrenheit, exhaustion sets in within fifteen to thirty minutes, and death occurs in thirty to ninety minutes, depending, of course, on the condition of the victims and how much clothing they are wearing. At 32.5 degrees Fahrenheit, death can occur in fifteen minutes or less.

Had the crew of the *Marine Electric* had insulated survival suits—and the time to get into them—they would almost certainly have survived for twelve hours and been rescued. Unfortunately, they did not have such protective gear.

The ship itself floated upside down for hours before finally disappearing from sight about 11 A.M., scattering her load of coal over the sea floor. Four days after the disaster, two Coast Guard divers (one of whom was the man who had been refused permission to help with the rescue) dived on the wreck to find if anyone was still alive inside her. The storm had passed, and the day was clear and sunny; the sea was calm. As one of the divers later commented, they had no difficulty finding the location because a Coast Guard buoy tender, dispatched to locate the wreck with its sonar, had become fouled in one of the *Marine Electric*'s long, buoyant, plastic mooring lines, which had floated up to the surface. The other end of the black-and-yellow line remained fast to the sunken ship, and the divers simply followed it down some 120 feet.

At the wreck, the divers encountered poor visibility—only about 10 feet—and strong currents that alternately pulled them under the hull and pushed them away. With the butts of their diving knives, they pounded on the hull on the very slim chance that the men trapped inside might still be clinging to life in a chamber partly filled with trapped air. There was no response. The divers had only enough air in their tanks for ten minutes of bottom time, so they surfaced without making further observations. When the Coast Guard Marine Board of Investigation met a few days later, the Coast Guard declined to send them down again as neutral observers, on the ground that the Coast Guard did not have decompression equipment and therefore could not permit its divers to go below 70 feet.

The Marine Board of Investigation of the United States Coast Guard convened in Portsmouth, Virginia, on February 16, 1983, to examine the case of the *Marine Electric*. Such investigations typically followed a sordid ritual. The shipping companies and their executives and their experts would testify that the ship had been in sound condition when it put out to sea. The Coast Guard would back the shipping companies with records of safety inspections (these were often fraudulent, a detail that was generally winked at). The classification societies, which also inspected the ships periodically so that they could get their insurance renewed, fell compliantly into line. Everyone then did their best to pin the blame on the captain and the other officers: they were the ones commanding the ship—if that ship wasn't safe, why did they take her out to sea?

The investigation into the cause of the sinking of the *Marine Electric* might well have taken this course had it not been for a few men: Gene Roberts, the editor of the *Philadelphia Inquirer*; his maritime reporter, Robert Frump; another reporter, Timothy Dwyer; and Captain Dominic Calicchio of the Marine Board of Investigation. Roberts believed that a large injustice was going to be perpetrated in the case of the *Marine Electric*. When he picked up the news of the sinking from the news wire, Roberts immediately assigned Frump and Dwyer to the story, promising them unlimited backing. Their investigative skill and devotion to the truth resulted in a story that changed the merchant marine system in America for the better and won them, their editor, and their newspaper an award. Much of the information here presented on the *Marine Electric* is based on Frump's and Dwyer's reportage in the *Philadelphia Inquirer* and on a book that Frump later wrote, titled *Until the Sea Shall Free Them*.

Captain Calicchio was one of the three officers who made up the Marine Board of Investigation. He had formerly served in the merchant marine himself before joining the Coast Guard and could appreciate the plight of mariners sailing on worn-out ships. He was an honest man in a system that rewarded "team players" rather than ethics. The other two board members were Captain Paul Lauridsen, the chairman, upright but disinclined to make waves, and Lieutenant-Commander E. F. Murphy, the recorder, a younger man who had not served on such a board previously. Two members of the National Transportation Safety Board sat in on the hearings and later issued their own report.

The investigation included testimony by the survivors and Captain Farnham as well as management from MTL, Coast Guard inspectors, and American Bureau of Shipping (ABS) surveyors. Direct evidence from the wreck, as interpreted by experts, also played an important role. Skilled divers went down numerous times to inspect the wreckage and take still photographs to document their findings. An ROV took still photos and videotape. Finally, divers secured samples of metal from the edge of the gap where 240 feet of the ship's midsection had been torn loose; these were tested by a metallurgical lab in Oklahoma.

The sequence of events underwater went as follows.

On February 16, the first team of Coast Guard divers went down and found the vessel lying upside down in about 120 feet of water on a sandy floor. There was about 75 feet of water above the wreck's upturned bottom. Since visibility was very poor, they saw little (in fact, they mistook the bow for the stern, where the engine-room crew was trapped) but were able to make out a break in the hull where the mooring line was attached. The same afternoon, the Coast Guard made a sonar scan and confirmed that the hull lay in two pieces.

Meanwhile, Marine Transport Lines was not idle. Seeking evidence to show its lack of culpability, the company sent seven divers to inspect the wreck between February 20 and 24. Before the divers went down, they conducted their own side-scan sonar survey for general orientation. Equipped with helmets and air hoses to extend their bottom time, they made two orientation dives before getting down to the serious work of looking for damage. The divers had no trouble finding their descent line, for the Coast Guarders had tied a life jacket to its upper end.

On the morning of February 22, one of the divers found the hole in the bow, which he described as half-moon shaped, 4 or 5 inches long, and 2 or 3 inches wide. (Apparently, the divers did not carry measuring tapes with them.) One diver took a 14-inch welding rod and probed the hole with it. He met with no resistance—evidently, Cusick's coffee-can-bottom patch and its cement-box backing had come loose. The diver then worked his way up and over the bow, finding the port anchor and the ship's name on the hull. He found a small horseshoe-shaped crack about 8 feet back from the port bow at the 27-foot draft mark. The starboard bow was buried in the sand beyond its railing. The diver saw no signs of damage on the foremost 10 to 15 feet of the starboard bow, nor did he see the starboard anchor, which would later play an important part in the hearing. He did not mention *not* having seen it, either.

The diver swam under the hull where a low spot in the sand gave a meager clearance and saw that the dry cargo hatch cover was missing. Hawsers were shooting out of the hole "like spaghetti," he said, and being sucked back in by the current every few seconds. He looked at the hatch dogs and found that one was threaded all the way up but had no nut on it; two others looked as if someone had sheared them off with a bolt cutter.

On the afternoon dive, another diver worked his way back along the ship's upturned bottom until he found a very large crack running clear across the hull, following the line of a weld. The metal appeared to be fractured in the area affected by the heat of welding. Using arm measurements, he later calculated the crack to be located about 38 feet aft of the forefoot of the bow. The crack was about 12 feet wide near its beginning, and it got wider as it progressed across to starboard. The hull plates were bent inward for about 2 feet on each side of the crack, and they had jagged edges. He followed the crack down for about 28 feet until he found himself inside a cavity. He telephoned his supervisor (the air hoses also carried communication wires) and learned that he was inside the starboard deep fuel tank. Although visibility inside the tank was only 3 feet, he explored the tank and found little or no evidence of distortion beyond a few feet from the edges of the crack. Instructed to look for abrasion marks on the ship's bottom (evidence of grounding), he found none. All the steel of the bottom looked to be fairly new, with no signs of deterioration. As for the hatch covers, none of the seven divers saw them.

Three more dives were made on February 24, ascertaining only that the stern was in line with the bow section but not quite as upside down. The stack, which had nearly hit Eugene Kelly, had been ripped off and lay 6 or 7 feet from the main deck.

Some months passed, and in mid-May a firm named Steadfast Marine made another sonar survey for MTL. This one produced a paper tracing that showed the hull was in one continuous piece, although 240 feet of midship section were missing. (Later a magnetometer survey indicated that a strip of the main deck, buried in the sand, still connected the bow and stern pieces.) The heading of the bow was meticulously recorded, for it could give a clue as to which way the ship was facing when she went down.

On May 24, the research vessel *G. W. Pierce*, chartered by MTL, sailed out to the wreck site, bearing not only a team of crack divers but also a 900-pound ROV with four thrusters, a video camera and a still camera, a manipulator, and lights. Its tether was 400 feet long, sufficient for the depth at which the *Marine Electric* lay. The divers discovered that the rudder and propeller had been made off with by unknown illegal salvagers who had blown them free with explosives and lifted them to the surface with air bags. Two bags were still tethered down below, ready for the next load of pirated salvage.

The ROV ran into a bit of trouble: a radio beacon placed on the wreck to orient it malfunctioned, and the ROV's magnetic compass was thrown into confusion by the wreck's steel hull. As a result, it was not always possible to give exact coordinates for the patch of hull being surveyed. The pilot tried to guide the ROV along the welding seams, but the strong currents made the job difficult. Still, the ROV managed to produce about twenty hours of videotape.

A diver was sent down to photograph the port anchor and its massive chain. He reported that both were intact and housed tightly in the anchor well. But the photos he took, plus eyewitness reports from another dive, revealed something disturbing: undetonated explosives were fastened to the anchor chain, presumably by the same pirates who had made off with the rudder and propeller.

Videos from the ROV showed that the transverse crack across the bow was larger than had been thought. Also, the bow showed pushed-in areas and cracks on both sides, while the stem had been fractured at the 21-foot mark and bent to starboard, indicating that the *Marine Electric* had struck the bottom bow-first. The starboard anchor and chain were nowhere to be found. Stolen? No one could say.

Divers using SCUBA were dispatched to investigate the starboard side of the main deck around the riding pawl of the anchor. They found two rusty links of chain inside the riding chock, with their ends disappearing in the sand. The turnbuckle—a tightening device familiar to every do-it-yourselfer—on the starboard devil's claw was attached to the pad-eye in the main deck, but one end was buried under the sand. The divers could not see the wildcat. On June 9, the divers investigated a large "sonar contact" a little more than 1 mile (about 1,800 yards) southwest of the main wreck. It was the missing midsection, sitting upright with most of its starboard side missing—it may have blown out when the boilers exploded. What was left of the interior was a jumbled and mutilated mess.

The R/V *G. W. Pierce,* apparently shuttling back and forth between the wreck site and the mainland 30 miles away, returned again on June 11 for a more detailed survey. The divers removed the starboard devil's claw, which MTL's lawyer thought would be a useful piece of evidence as the Coast Guard hearings continued. The turnbuckle for the devil's claw was found to be extended to almost the full length of the barrel, which is the part that houses the two threaded ends of whatever it is supposed to pull tight. That is to say, it did not appear to be completely tightened. Three or four turns, said the Coast Guard report, would have disconnected it entirely. Only two or three threads were visible inside the barrel, but they showed no signs of being stripped. There was damage, though. Three inches of the tip of the two-pronged devil's claw were broken off, and the bolt that held the turnbuckle to the deck was bent as if from severe tension on the devil's claw. (Considering that each anchor weighed 8 tons and the metal of the chain's links was 2 5/8 inches in diameter, there would have been considerable stress indeed if the anchor had come loose.)

On June 20, the *Pierce* brought the ROV and the divers back to the site. This time the divers found the starboard anchor chain and recovered 17 of its links, burning it free at the end next to the ship with an underwater torch. The length that was recovered lay outboard and forward of the chain stopper assembly. The divers tried to follow up the chain between the riding pawl and the windlass, but the sand was too much for them to contend with.

On June 25, the divers took a powerful water-jet down with them and attempted to blast the sand clear to look for the rest of the starboard anchor chain, but they could not find it. The ROV was taken out to the location where the ship was abandoned and also to a point half a mile northeast of the wreck, but it, too, found nothing. The starboard anchor and the remainder of its chain were never found.

At this point, a private diver named Jeremiah Shastid enters the story. Shastid, an ex-marine diver and Vietnam veteran, wanted to establish himself in the commercial diving business. To do this, he needed some successful assignments on his record. His first job offer came from lawyers representing the families of some of the lost crew members, who scraped together about $4,000 to cover his expenses. The seamen's union, too, contributed a small donation.[14]

Shastid's outfit could be described as barer than bare bones. He acquired a secondhand open skiff and a big outboard motor to run it. His SCUBA gear he presumably had already. He got an underwater camera and high-speed film. Paper for his notes, including the map coordinates of the wreck, he put in a zip-lock bag. One could hardly be blamed for thinking the operation amateurish. But Shastid got some very professional results despite setbacks. As Dwyer told it in the *Inquirer*, one day in April Shastid and his helper zoomed out from the marina at Chincoteague Island to the marker buoy 30 miles away, only to be confronted by a clam boat from Maryland, out to help itself to whatever it could pry loose from the wreck (scrap always

---

[14] This is according to Frump and Dwyer. The Coast Guard says that Shastid was hired by "union interests."

has a market value). Its occupants brazenly flew the traditional Jolly Roger flag of pirates. The pirate crew menaced Shastid with automatic rifles and ordered him away. Unarmed, Shastid followed the prudent course of retreat—but he was not cowed, even though his assistant quit on the spot. Shastid engaged a new helper, a seventeen-year-old who lived in the same cheap motel in which Shastid was lodging.

Shastid returned to the wreck in May, bringing with him not only his young assistant but also the reporter Timothy Dwyer. As his assistant attempted to make fast to a buoy, Shastid was hailed from the bridge of the *Pierce* by a vice president of MTL, who demanded to know what he was doing. "Diving," replied Shastid.

"Not today you're not," called someone else from the research vessel, adding that they were about to drop four 2-ton anchors and didn't want to hit Shastid with one. The threat only made Shastid more determined.

Lieutenant-Commander E. F. Murphy, aboard the *Pierce* as a Coast Guard observer, asked where the men's life jackets were. Shastid rummaged around and finally produced two—one short of Coast Guard regulations. Murphy informed the ex-marine that he was in violation of basic Coast Guard requirements for small boats. Then Murphy recognized Dwyer, whom he had seen at the Marine Board of Investigation hearings. He asked the journalist what it would look like if a reporter from the *Philadelphia Inquirer*, who was investigating the Coast Guard for allegedly permitting unsafe ships to put out to sea, were arrested for being out at sea himself in an unsafe boat without a life jacket. Dwyer turned the question back to Murphy.

The Coast Guard officer and the MTL executives considered the situation for a moment and realized that arresting Dwyer would make them look very bad in the eyes of the public. So instead they forced themselves to be courteous and agreed to let Shastid dive for ten minutes. Down below, Shastid found the pirates' two air bags. He cut them both loose, intending to take them for himself as payback for the pirates' threats. First he tried to tow them, but his boat could not move them. They were too heavy for him to lift into the boat, muscular though he was. (Even if he had succeeded, he would have had a difficult time keeping them from tumbling off on the return trip since they were bigger than his boat.) A second attempt at towing failed. Then one of the bags filled with water and sank. A third attempt at towing the remaining bag failed when the boat still couldn't move with it attached. Exasperated, he slashed the bag with his diving knife, saying that, if he couldn't have it, nobody could. Smiling for the benefit of the onlookers aboard the *Pierce*, he added that he would not have taken such measures if the pirates hadn't brought out their guns and threatened him.

The following day, MTL, evidently trying to ingratiate itself with the press, invited a group of reporters to visit the *Pierce*. Shastid volunteered to bring them out in his runabout. Once aboard, the reporters gravitated to the viewing room, where people were watching the progress of the ROV. Shastid, curious and interested, joined them. Among the crowd was MTL's top lawyer, who flew into a rage and ordered Shastid out of the room. After the two men had indulged in a shouting match, the lawyer consented to give Shastid lunch but *not* to let him watch the screen.

Undeterred by his lack of information from MTL's ROV, Shastid came back to the wreck site a number of times, making thirty-eight dives in all and documenting his findings with photographs. Each dive lasted about fifty minutes, pushing the limits of safety. He succeeded in photographing the hatch covers, which the other divers and the ROV had failed to do because of the difficulty in reaching them. When he testified before the Marine Board of Investigation, he compared the damage he had observed to that which would have been caused by an explosion. Of the 300 shipwrecks he had dived on, he said, the *Marine Electric* was the most badly damaged.

Meanwhile, the shipping company had a tremendous financial stake in the outcome of the investigation since it was already being sued by the families of the dead crew. Under maritime law, as Dwyer pointed out in one of his articles, if MTL were found not liable for having sent an unsafe ship to sea, its damages would be limited to the value of the ship itself, estimated at $10,000,000, plus the value of the cargo, which was $2,000,000.[15] If, on the other hand, MTL were found liable, each claimant would be entitled to a share limited only

[15] This figure is taken from the NTSB report.

by the assets of the company, assuming that a jury found their cases valid. The total of the claims was in the neighborhood of $100,000,000; at the time, MTL's market value was a bit less than one-fifth that amount. Thus, the company had an extremely powerful incentive to establish its blamelessness.

As mentioned earlier, it was customary for shipowners to place the blame on the officers of the lost ship, with the captain first in line. But Captain Farnham had been on vacation when the *Marine Electric* went under and was obviously in the clear. The relief master, Captain Corl, had died with his ship, and his body had been lost without a trace. Thus, it was the chief mate, Robert Cusick, who was next in the line of fire, and MTL's lawyers were prepared to do everything they could to destroy his reputation as a seaman.

First, however, a theory had to be dealt with, one that most mariners would instinctively jump to. The water was fairly shallow in the area where the *Marine Electric* had stood by to help the *Theodora*, with shoals as close to the surface as 40 feet in calm weather. The *Marine Electric* was drawing 34 feet, and the waves were 20 to 25 feet high, with possible rogue waves of 44 feet. It seemed logical to conclude that the *Marine Electric* had bottomed on one of the shoals in the trough of a wave, sustaining a crack that spread to fatal proportions. The captain and the crew, preoccupied with fighting the storm, might not even have noticed the impact.

It was a perfect story for the shipping company: the brave and selfless crew, in their safe, though elderly, ship, sacrificed themselves for fellow seafarers in distress; all became victims of one of the worst East Coast storms in decades. Everyone came out looking good. And the giant crack that the divers had found near the bow seemed to bear the theory out. But was it true?

Dewey and Kelly and Cusick were summoned from their homes near Boston. Dewey was the first to testify. Marine Transport Lines had excellent lawyers, and the lawyers were ruthless in their determination to skewer the crew to preserve the company's guiltlessness. Dewey, a mere seaman, and Kelly, the relief third mate, were not significant players in this courtroom drama. It was Cusick who was assigned the starring role.

The grounding theory was dismissed by the Marine Board of

Investigation when the captain of the *Theodora*, which had a depth finder, testified that the water where the *Marine Electric* had stopped to safeguard him was never less than 110 feet deep. Cusick testified that his ship had never entered water less than 90 feet deep and that at no point had he felt her strike bottom. The Coast Guard itself estimated that the *Marine Electric*'s closest approach to any shoals that were on the map was at least 3 miles distant, and the water was never less than 12 fathoms (72 feet) deep.

With the heroic theory destroyed, the company lawyers were forced to zero in on Cusick's culpability. First they attempted to show that he had neglected to make sure that the starboard anchor was properly secured. In the storm, they claimed, it had come loose and flailed about, punching a series of holes in the ship's hull. Through these holes, the water had poured in and sent the ship down by the bow. As evidence, they produced the turnbuckle of the devil's claw and some images that showed the alleged holes.

Calicchio disproved this scenario on two counts. He pointed out it was possible, even probable, that the anchor could have come loose once the ship began to sink, but then it would have had to swing *up* instead of *down* to make those holes, defying the law of gravity. The MTL's marine architect was compelled to admit as much. Then, with a series of simple questions, Calicchio got the head of MTL's underwater investigative team to concede that the turnbuckle did not have to be screwed completely closed to hold the anchor chain taut. The holding power of the turnbuckle depended entirely on the few threads that engaged the threads of the barrel, no matter how much or how little of the threaded rods protruded beyond them.

The MTL's lawyers then badgered Cusick about the condition of the hatch covers, on which Cusick had blamed the flooding of the cargo holds and the subsequent capsizing. As chief mate, was he not, they asked, responsible for keeping them in good condition? Cusick replied that he had told Captain Farnham about the holes and cracks and that Farnham had not only told him to keep patching them with epoxy and tape but also asked him to sketch a plan of the worst areas. This Cusick had done. As he testified, he was not authorized to carry the matter further.

During the hearing, many embarrassing pieces of information were revealed. When the *Marine Electric* had gone for her scheduled repair work at a Jacksonville shipyard, thirty-one steel patches, or "doublers," had been welded over the worst holes in the hatch covers, but the remainder of their surfaces had not been taken care of. Cusick told the board that he repeatedly had the crew scrape the flaking rust off the hatch covers and give them several coats of red lead (a traditional rust preventative) before coating them with paint, but finally the covers were so deteriorated that he gave up and simply had the crew paint them with a mixture of fish oil and hull black.

Third Mate Kelly told the board that he had knocked off sheets of scale up to 3 × 3 feet from some of the hatch covers with his knife. All the hatches had holes ranging from pinhole- to fist-size. Some of the gaskets had not been fastened to the hatch covers and were hanging down loosely. Parts of the track on which the hatch covers ran were missing or bent, due to damage by the heavy grab bucket with which the coal was unloaded. The hatch covers would jump their tracks routinely when being opened or closed, and the crew had to break out chain falls and come-alongs (a kind of ratchet-lever device) to get them back on. In Kelly's opinion, the constant work of grinding corrosion off the hatch panels and then painting them was no more than camouflage. Moreover, when the hatch covers had been repaired in Jacksonville, they had been removed from the ship and not brought back until the day before the *Marine Electric* was scheduled to sail. They were still warped and would not open or close properly. They had to be forced back onto their tracks. And the replacement rubber gaskets with which they were fitted were too short and did not seal properly.

Maxwell Graham, a representative of the MacGregor Company, maker of the hatch covers, had come to check the hatch covers *after* the Coast Guard inspectors and the ABS surveyor had approved the work. He reported the same deteriorated conditions that Cusick had testified to and recommended that several panels be replaced. Joseph Thelgie, fleet director of MTL, sneered that the man was simply a salesman attempting to peddle his wares. The recommended replacements were not installed.

A surveyor from the ABS had visited the *Marine Electric* in Baltimore on February 24, 1982, and walked around the deck with Cusick for about half an hour, looking at the condition of the main deck and the hatches. Cusick testified that he pointed out numerous epoxy patches, taped-over holes, and doubler plates, but the surveyor did not comment on them and did not take any "appropriate action." At the hearing, this surveyor could not recall any deficiencies in the hatches. His report graded the hatches as satisfactory, and he also gave a satisfactory mark to the cargo holds, although he never entered them. Of course, he did not see the steel plates over the bilge pump intakes since he did not go anywhere near them. The surveyor recommended that the vessel "be retained in class"—that is, he deemed it in good enough condition to have its insurance renewed.

Another ABS surveyor, Serge Simeonidis, testified that he had carefully checked the condition of the hatch covers during the ship's dry-docking in Jacksonville in February 1981. He did this, he said, on February 22. However, the aforesaid hatch covers were not on the ship and were not returned until February 23, as Cusick had previously testified. Under questioning by Cusick's lawyer, Simeonidis asserted that he had visited the *Marine Electric* about thirty times during this stay in the dry dock; on one visit, he had stayed seven hours. He said that he had never noticed whether the hatch covers were on or off the ship, which contradicted his claim that he had checked them carefully. He also said that he had looked at the hatch covers from the inside and the outside. If he had thought the hatch covers were in bad shape, he said, he would not have certified them.

A former surveyor from the ABS, one Richard Conway, had worked for the bureau from 1977 to 1981. He had inspected the *Marine Electric* in February 1980; he testified that the hatch covers at that time were "a wreck." He said that he recommended that repairs be made, but MTL would not allow him to inspect the repairs. "The hatches were never closed for me," he said. "The owners never closed them for me. They were in a hurry to get out of there. I was costing them a lot of money." He did not notify the Coast Guard of the deficiencies because a Coast Guard inspector was aboard the ship at the same time. In any case, it was not his practice to report deficiencies to the Coast Guard because that was not his duty. "I notify the bureau, and it's up to them to notify the Coast Guard," he said. It may be noted that the ABS surveyors were paid directly by the shipowners whose vessels they were inspecting. Conway followed this admission with a more startling one: "If I had [reported the deficiencies to the Coast Guard], I would have been fired. No one would have let me on their ship."

Questioned by Calicchio, Conway acknowledged that he had signed a report stating that the hatch covers were satisfactory. He excused himself by saying that he knew the ship was scheduled for a regular inspection the next year. That was the one performed by the casual Mr. Simeonidis, who never even looked at the hatch covers.

Another ship surveyor told the reporter from the *Philadelphia Inquirer* that Conway was a good inspector, adding, "But they don't like whistle-blowers in this business." He went on to say that, if one ABS field office got a reputation for being strict, the shipowners would simply shift their business to another port.

James Guidish, a former Coast Guard inspector who had moved to the ABS, found himself in disagreement with Maxwell Graham of the MacGregor Hatch Company, who testified before him on March 22. Graham said that he had visited the *Marine Electric* four times after February 1981 to check the hatch covers and found each time that they were not watertight. He had repeatedly sent letters to MTL, urging it to replace the hatch covers or make extensive repairs to them. Thelgie had previously stated he had received only the last letter from Graham. The result of Graham's warnings was that Thelgie finally spent some company cash to order one new section of the hatch covers.

When poor Guidish came to the witness stand, he admitted that he had never inspected or tested the hatch covers as Coast Guard regulations required. He was under the impression that they had been repaired while they were off the ship. Calicchio asked Guidish, "Are the hatches supposed to be watertight according to regulations?"

"Yes, sir," was the answer.

"Were they watertight when they left the shipyard?"

"I am sure they were."

"How do you qualify that answer?"

"I believe that the owner's port engineer made arrangements to test the hatches."

"Did you yourself test the hatches?"

"No, sir. I don't think I know how to test a hatch."

Guidish said that he had not tested the hatch covers because he assumed that the company would not let the ship go to sea if they were not watertight. He ended with a classic example of circular reasoning: "The ship had to go with tight hatch covers. The owners would be foolish not to have a ship with tight hatch covers."

Basil Andriopoulos was MTL's marine superintendent for the *Marine Electric*, the man in charge of her maintenance. Andriopoulos said that he had visited the ship on February 8 in Massachusetts, while she was unloading coal, and could not inspect the hatch covers properly because they were open. He asked the chief mate to make sketches of the deteriorated areas, he said. Thelgie said he had seen the sketches—but he apparently deferred taking action on them. Andriopoulos, under questioning, admitted that he had not checked the hatch panels. He also told the board that MTL had requested the postponement of the scheduled inspection until April because the customer, the New England Power Company, did not want to run out of coal and asked that the shipment not be delayed. (The power company said that it had never made such a request and, furthermore, had an excess supply of coal on hand.)

Andriopoulos embarrassed himself and his employer further when he contradicted himself about whether or not he had authorized the use of patches over the holes: first he said "never," and then he said that he did authorize their use. He admitted that he did not know if the hatch covers had been tested for water tightness after the alleged repairs were made. "I assume they were," he said. When asked if the patches met the strength requirements of the Coast Guard and the ABS, he replied that they did, but he confessed that he did not know what those requirements were or whether the patches had ever been tested for strength. He frequently answered questions with "I don't remember."

The investigators also learned that, when the *Marine Electric* made her trans-Atlantic grain voyages, the joints between the hatch covers were simply covered with a tar-paper-like product called

Ramneck, topped off with roofing tar. The Coast Guard and the ABS were never notified of this economical expedient.

The ability to pump water out of the bilge is vital to keeping a ship afloat. But on previous voyages, the bilge pumps of the *Marine Electric* had become clogged with coal dust from the cargo. The solution? Steel plates were placed over the drains to keep the coal dust out—so ordered by the permanent master despite his insistence that all such matters were the province of the chief engineer. When asked how the bilges could be pumped under such conditions, Cusick could only shake his head mutely.

Had the chief mate made sure that the hatches were watertight after loading was completed? He replied that the hatches had not fitted properly since they were repaired at the *Marine Electric*'s last dry-docking in 1981. The rubber gaskets that were supposed to seal them were too short. Some of the dogs that fastened them down were not functional. Had he reported this to the master of the vessel? Yes. It was not his responsibility, nor did he have the authority, to take the complaints further. (In fact, he would almost certainly have been fired had he done so.)

Had he notified the captain of the holes and cracks in the hatch covers and decks? Yes, in fact Captain Farnham had asked Cusick to sketch the damage for future repairs. He had done so and turned the sketches over to the captain. What had been done with them after that he did not know. But the repairs had not been made.

The situation looked very unfavorable for Marine Transport Lines. But the cross-examination, still to come, would give the company's lawyer a second chance to discredit Cusick and save his client from substantial damage suits by the families of the deceased crew and officers. Just before the cross-examination, Cusick was called back to Boston by his cardiologist; blood tests done while he was still in the hospital recovering from hypothermia showed signs of danger. The hearing was accordingly postponed.

Covering the story, reporter Robert Frump pointed out that MTL was not simply being greedy and callous in its attempt to clear its name. Shipping companies were constantly being victimized by fraudulent damage suits from seamen, longshoremen, and other employees who claimed that they had been injured on the job. Juries tended reflexively to side with the "little guys" against the powerful and wealthy corporations. If the corporations did not resist these suits, they would be drained of capital and go out of business. The whole system bred cynicism and mistrust.

The *Philadelphia Inquirer* published Frump and Dwyer's story on May 2, 1983, on the front page. The banner headline read "DEATH SHIPS." It recounted in chilling detail the sinking of the *Marine Electric* and the struggles of her crew in the sea. It was an emotionally powerful and excellently researched piece. The sequel, on May 3, revealed an unsavory picture of collusion between the entities involved in American merchant shipping. The feature detailed a Faustian bargain between, primarily, the shipbuilding industry, which wanted to sell ships to the shipping lines; the shipping lines, which needed the business of carrying cargoes; and the seamen's unions, which needed jobs for their members. The three groups exerted a great deal of power in Washington, not least through campaign contributions.

A well-intentioned piece of legislation, the Merchant Marine Act of 1970, provided generous subsidies to American shipping lines and also specified that a certain percentage of American products each year had to be shipped in American-built ships with American crews. Unfortunately, the cost of new ships built in American yards increased to the point where American shipping lines could not afford to buy them. In order to meet the law's requirements, the shipping lines had to rely on the older ships in their fleets. To maintain profits, they typically skimped on maintenance. The Coast Guard had strict and well-enforced safety standards for the newer ships, but the old ships were exempted from these rules. The result was what Frump called "serial sinkers": ships that sank one after another due to failure of the hatch covers or the hull, which sometimes broke clean in two. Many seafarers died as a result of these failures.

Cusick's attorney withheld Shastid's photos until the end of the hearing, when an MTL witness was defending the condition of the hatch covers. The lawyer had the witness examine a series of pictures; among them were some of Shastid's most graphic photos. They showed the deficient hatch-cover gaskets and the deteriorated condition of the covers themselves. The witness was forced to concede the point.

The Coast Guard also performed extensive and highly technical calculations of the stability the *Marine Electric* would have had under various conditions, and in September 1983 it sent divers to secure metal samples from the fractured area at the end of the bow section, aft of the No. 2 cargo hold. The samples were taken from plates selected after the Coast Guard reviewed videotapes and chose the plates that could be identified with the greatest certainty. Metallurgists at the National Bureau of Standards (an agency later abolished under a Republican president) subjected the samples to something called the Charpy V-notch impact test, which involves notching them, laying them on a support, and having them struck by a weighted pendulum to break them. By comparing the position of the pendulum at the end of its stroke to its position at the beginning, the ductility of the metal can be calculated. Other samples were tested for chemical composition. All were found to be within the safety guidelines. This was to be expected since much of the bottom plating had been replaced with new steel within the two years before the sinking. Three plates taken from different sections of the bottom near the keel were checked for thickness and found to be a little thinner than the design of the ship called for. One showed signs of pitting from corrosion by the salt water.

The NTSB, conducting its own investigation, released its report in March 1984. Among numerous conclusions was the determination that the *Marine Electric* had not struck ground but had sunk because of flooding from an "undetermined structural failure." The No. 1 hatch cover had indeed failed but not until flooding had caused a significant trim by the bow. The hatch covers, said the report, had "probably" been weathertight when the ship left Norfolk. The great loss of life was attributed to the lack of survival suits and a life-raft design that made the rafts difficult for persons in the water to climb into.

The report came down hard on the Coast Guard and the ABS for their laxity in inspections; its severity was almost unprecedented. Some of its recommendations to the Coast Guard went unheeded, even the one requiring survival suits for all personnel on board a vessel operating in water where the temperature is below 60 degrees Fahrenheit. Other recommendations were followed at a later date.

The Coast Guard, apparently fearful of opening a can of worms, temporized with its own report, which it did not release until January 1985. According to Frump, it was embarrassed into releasing it by the head of the Marine Engineers' Benevolent Association, a powerful maritime union, who ordered 5,000 copies of the main *Philadelphia Inquirer* exposé and mailed them to government officials, union officials, and shipping-line executives.

Late as it was, the Coast Guard report was frank about the deficiencies of Coast Guard inspections and made recommendations to remedy them. It also took the ABS to task, saying, "Basically, ABS surveys are oriented toward protecting the best interest of marine insurance underwriters, and not for the enforcement of Federal safety statutes and regulations." It also noted that the ABS surveyors were subject to the influence of the shipowners or other interested parties. Paragraph 44 of the report's general conclusions said, in part, "The assertion on the part of the surveyors in their written reports and their testimony before the Board ... raises questions about the professional integrity of their surveys." (A later review by the Coast Guard commandant softened this conclusion, saying that the ABS should not be blamed for the actions of individual surveyors but that there was a "need for more formal oversight of surveyors' performance.")

About the owners' culpability, the report commented, "The ship was poorly managed and horribly maintained with repairs to the hatch covers, main deck, and holes in the cargo hold area...." The report also attributed the sinking to the flooding of the first two cargo holds when the hatch covers gave way under the pounding of the sea.

Most satisfying of all, from the viewpoint of the three survivors and the families of the dead crew, the Coast Guard recommended that Fleet Superintendent Thelgie and Captain Farnham be prosecuted for criminal negligence. MTL's lawyers were able to avert this threat, but the company's reputation suffered, and the warning to shipowners was plain. According to Frump, within the next two years, more than seventy overaged and unsound American bulk carriers were sent to the scrap yard. In 1986, a U.S. attorney in the Norfolk office prosecuted MTL for failing to report the hole in the hull that Cusick had patched with epoxy and a coffee-can bottom—a felony under federal law. MTL was eventually convicted and fined $10,000, a ridiculously small sum but symbolically very important. The Coast Guard also fined MTL $10,000.

Marine Transport Lines struggled on for a few years, facing high operating costs, and was eventually swallowed up by another corporation.

* * *

The *Rema*, like the *Marine Electric*, was not a glamorous boat. Merely a rust-splotched coastal bulk carrier, she was carrying an unglamorous cargo in April 1998: crushed redstone of two sizes—2 millimeters and 8 millimeters. Crushed rock of this size would probably have been used as aggregate in concrete or as a component in ornamental paths.

Unlike the all-American *Marine Electric*, the *Rema* was an example of multinationalism in action. She was built in the Netherlands in 1975, was owned by a shipping company based in the Bahamas, was operated out of the United Kingdom, and was registered in Belize, a small Central American country formerly known as British Honduras. She was nowhere near the size of the decrepit American collier, measuring a mere 195 feet in length and 31 feet in breadth and drawing only 10.25 feet fully loaded. When she sank in the North Sea in April 1998, four crew were drowned—not thirty-one as in the case of the *Marine Electric*.

Before her unfortunate ending, the *Rema* had carried a variety of cargoes, from grain to fertilizer to brewers' malt to stone. She shuttled between various ports in the United Kingdom and northern Europe, and on her last trip she was bound from Berwick-upon-Tweed, on the border between England and Scotland, to the minor Dutch port of Terneuzen—an unremarkable journey. She arrived at Buckie, in northern Scotland, on April 21, 1998, with a cargo of malt from Belgium. After unloading, the crew washed out the cargo holds in preparation for the next load, whatever it might be. According to the Marine Accident Investigation Branch (MAIB) report, the *Rema* enjoyed a good reputation for cleanliness and water tightness, so much so that a number of shippers specifically requested her services.

The crew, however, was another matter. When the *Rema* was checked out in Bremerhaven, Germany, in March 1997, the German Lloyd inspector had found ten deficiencies. One of them was substandard certificates of competency for the skipper and mate. Others included hull damage that impaired her seaworthiness, defects in her steering gear, plus smaller problems that an efficient crew should have taken care of. She was detained in port until these problems were fixed, with the exception of the certificates of competency. These subpar qualifications caused trouble again in February 1998, when the ship was held up for correction of deficiencies at Montrose, Scotland, for five days.

Under the 1974 Safety of Life at Sea (SOLAS), the *Rema* should have carried two lifeboats, although an equivalent was acceptable. What the *Rema* had instead were two life rafts with an awkward launching arrangement, plus a man-overboard rescue craft. These devices were deemed acceptable by the authorities in the Netherlands, where the twenty-three-year-old carrier was built, and later approved by Belize.

The Scottish inspector at Montrose finally authorized the *Rema* to go to sea under two conditions: that she not go more than 100 miles from the nearest land, and that there should be a life raft for each side of the ship, large enough to carry everyone on board. Since the crew was five men strong at its maximum, the last condition was not a problem. There was also the matter of not having two qualified radio operators. But since Belize insisted that the master and the mate were qualified officers, the radio-operator requirement was waived.

The master, also called the skipper in British usage, was forty-one-year-old Michael Stuart Clayton, who had twenty-eight years of experience at sea. The mate was his brother Robert, aged forty, who had a mixed record of experience as a welder/boilermaker and boatyard manager ashore and as a relief-mate-cum-skipper aboard coastal vessels. He had also captained sailing yachts. Michael Clayton held a Belizean master's license for coastal vessels up to 1,800 gross tons, and Robert held a recently issued Belizean mate's license. Neither of the brothers had a British certificate of competency.

The crew was supposed to include two able seamen (an advanced grade of seaman who has had special training) and an ordinary seaman. The *Rema*'s two seamen were Andrew Richard James, aged thirty-four, who had never been to sea before, and Shaun Norton, aged twenty-six, who had been sailing sporadically on the *Rema* for about a year. Neither had taken any training courses. And the crew was one man short. (There were originally five men, but one simply walked off the job in port for "personal reasons." There was probably not time to replace him if the *Rema* were to make her schedule.)

The MAIB report dryly notes that, while the master had originally selected his crew on the basis of their seafaring experience, later on the men were picked on the basis of availability and family connections. It should also be noted that Skipper Clayton was the actual owner of the *Rema*—he had purchased her from Dutch owners in 1995—which might have given him an incentive to skimp on maintenance. Apparently, the Bahamian company listed as the titular owner was a front behind which Clayton operated.

The *Rema*, in the course of her working life, had suffered accidents. One of the more recent occurred in March 1997, when she lost electrical power and steering in a German port and collided with a bridge, which caused the damage that had to be repaired in Bremerhaven. More repairs had to be made in Britain; the master and crew, with a few self-employed helpers, took care of them. Only a month before she sank, the *Rema* collided with a concrete jetty and suffered a small hole in the forepeak tank and a 6- to 8-inch split in the side of the deep tank. Again the crew fixed the damage with their welding torches; the hole was repaired by simply welding a bolt head into it. Apparently, she had also kissed some rocks with her propeller at some point, for when she was in dry dock in 1997 the tips of her propeller blades were obviously either broken off or bent over. The master said he knew about the damage and would have the propeller repaired at his vessel's next scheduled dry-docking, in June or July 1998. Michael Clayton was reputed to avoid reporting accidents and mechanical problems, preferring to repair them discreetly on his own. He was also known as a "rock-dodger," or one who liked to operate in shallow water or close to shore.

The *Rema* left Berwick-upon-Tweed uneventfully at 12:30 P.M.

on April 24 and headed for the Netherlands. The wind and seas were moderate. The rust-blotched and dented old coaster chugged south along the Yorkshire coast at a little over 6 knots (estimated from the damaged condition of her single propeller). A little after 10 P.M., a man who lived on the coast saw what he thought was a small coaster of an old-fashioned design heading first southwest, toward the coast, and then south by southeast, as if it were threading its way among the rocks, reefs, and shoals that line that stretch of the British coast. The ship's lights showed up clearly in the moonless night. This person thought it might have been the *Rema*, but he could not be sure. He did tell the authorities that he was afraid the ship would strike the dangerous Fairy Rocks and "spring a plate." (It turned out that this observer was mistaken. It was a different boat that he saw.)

Another possible witness, the Honorable Secretary of the Seahouses Lifeboat, said that he had seen a vessel motoring through a channel in the Farne Islands in midafternoon, between 3:30 and 4:30 P.M. Through his binoculars, he could see an *R* on its stern.

At 3:21 the next morning, the *Rema* sent out three *Mayday* calls, the last one incomplete. The master included the vessel's position, about 20 miles east of his usual route to the Continent. Presumably, he was agitated by the fact that his vessel was sinking under him, for his voice was rising as the last *Mayday* was cut off. The Humber Coast Guard station responded immediately but got only silence on the other end.

A helicopter reached the scene at 4:31 A.M., followed by two lifeboats. The rescue team found an overturned Zodiac inflatable, a life raft, a life raft container, five lifebelts, one life jacket, two survival suits, a thermal bag and a safety bag, and wooden debris—but no people. A large oil slick was spread over the water. In all, eight vessels of varying sizes took part in the search, working outward from the center point, the spot where it was presumed the *Rema* sank, until the search was called off at 9 P.M.

The accident got a lot of play in the local press, and public opinion demanded an investigation. A search vessel equipped with sonar and a depth sounder began work on April 26 and found the wreck. Two months later a survey vessel carried an ROV to the site for a detailed inspection of the wreck. This ROV was armed with a charged coupled device color camera, a silicon intensified video camera, two 150-watt lamps, and an obstacle-avoidance sonar to detect possibly dangerous objects that the lights could not pick up.

The corpse of the *Rema* was found sitting upright on the sloping sea floor at a depth of 197 to 213 feet, cushioned by a bed of heavy mud and shells. The wreck was in an area with strong subsurface currents, so the ROV was sent down only at slack-tide periods. The ship's cargo had been dumped on the sea floor under and around the wreck, extending out for about 250 feet on the starboard side.

The ROV was maneuvered to cover the hull from main deck level to bilge line and from bow to stern. On the port side, it found no signs of ruptures or breaches in the steel of the hull, but on the starboard side it sent back images of a deep V-crease in the plates below the wing of the wheelhouse and one porthole that had lost its glass. The stem was pushed backward, so that the whole bow was buckled. The wheelhouse was crushed so severely that the ROV could not get inside to survey the interior, and the starboard side of the stern was deeply creased. The galley and the crew accommodations had been so distorted and filled with loose equipment that nothing could be seen.

Although the rudder did not appear to be damaged, the propeller was in sad shape: three of its four blades were missing their tips, and the fourth was bent out of shape. All the hatch covers were in place, indicating that there had been no damage from waves, but most of them had suffered distortion, and the No. 1 hatch had been knocked so far out of line that its left rear corner was sticking down into the hold it had once protected. The port life raft was missing (it must have been the one that the original search-and-rescue team found), while the starboard raft had been released but was jammed behind its cradle. Tellingly, the EPIRB was still in its holder; it had never been released.

The pictures taken by the ROV showed what looked like a bad vertical crack in the hull plating. However, when the picture was corrected, the crack turned out to be a dark line in the paintwork. And spots that looked like severe corrosion were determined to be no more than discoloration. Some pitting was found, but it did not penetrate the steel of the hull. In short, the *Rema* was not nearly in bad enough condition to have sunk as a result of hull failure. A possibility remained that the master did not replace a leaky oil seal on the propeller shaft as he had said he had done before the ship sailed. But he would not likely have taken such a risk.

The MAIB undertook an exhaustive analysis of the sinking, perhaps because of rumors that it had been a planned insurance fraud. The wife of one of the missing crew members supposedly told her mother on the day after the sinking that her husband had phoned her and told her not to worry—the *Rema* was going to be sunk at a suitable time and place. But some detective work involving the telephone conversations between the crew and their relatives before the accident, and among the relatives themselves both before and after, turned up nothing to indicate that the April 25 sinking had been planned.

The MAIB and the Salvage Association worked up four different flooding scenarios and came to the conclusion that the *Rema* was the victim of a slow leak into the hold that let her proceed on an even keel with nobody noticing that she was settling ever lower into the water—until the moment before the *Mayday* call, when she precipitously sank bow-first. The nature of the damage to the hull indicated that this was indeed the case. Could the cargo have shifted when it got wet and unbalanced the vessel? This would have been a plausible scenario with other types of loose bulk cargo, but crushed stone has a fairly steep angle of repose and does not liquefy and turn to slurry readily. Could the ballast tanks in the double bottom have filled until they exceeded the vessel's buoyancy? No. Calculations showed that the *Rema* would have stayed afloat under these conditions. And the forward ballast tanks had imploded as she sank, which showed that they had been empty. The most likely scenario was that the vessel had hit the sea floor somewhere in the Farne Islands without the crew noticing; the impact had ripped a hole in her bottom, which slowly flooded the hold. The mate on watch at the time of sinking would have suddenly noticed the ship's dangerous condition and woken his brother, the master, who gave the *Mayday* call. The *Rema* then sank so fast that no one had time to get off. However, the ship's bottom could not be examined because it was settled into the sea floor.

The experts calculated that it took an extra 757 long tons of seawater penetrating her hold to sink the aging coastal freighter. But how the seawater got there no one could say. The MAIB did determine, however, that, if the vessel had had automatic bilge alarms, the crew could have started the pumps and kept her afloat. It recommended strongly that all cargo vessels of the *Rema*'s type, with one large cargo space not divided by bulkheads, be fitted with bilge alarms. And the chief inspector of Marine Accidents recommended that all vessels be fitted with a voyage data recorder (VDR), a device analogous to the black boxes installed on aircraft that collects and records onboard information such as navigational and engine data. Had the *Rema* had a VDR, it would have helped solve the mystery of her sinking.

* * *

The *Marine Electric* went down in a violent storm; the *Rema* slipped to her doom in calm seas. The American coal carrier was far larger than the British coaster, and her loss of life was more than eight times as great. Yet, despite these disparities, the two tragedies have a factor in common: neglect of necessary maintenance resulted in disaster for both.

5 Unsolved Mysteries: The *Estonia* and the *Edmund Fitzgerald*

The tale of the *Estonia* could well begin with the hallowed formula "It was a dark and stormy night." It was indeed a dark and stormy night when the giant car-and-passenger ferry ran into rough water on its scheduled run between Tallinn, the Estonian capital, and Stockholm, Sweden. Aboard were 989 passengers and crew plus a substantial load of cars, big trucks, and buses.

In the years after World War II, the trans-Baltic ferries had evolved from a means of basic transportation into floating tax-free shopping malls that included bars and restaurants and cabins of varying degrees of comfort and luxury. Towering 100 feet or more above the water, they were like floating resort communities. The palatial craft were especially popular with Swedes, who liked to take weekend trips on them: a ship would depart one evening and spend the night at sea; it arrived in Tallinn (or Helsinki or one of several other ports) the next morning, with the day free for sightseeing or more shopping; and it returned to Stockholm the following morning. At certain times of the day, one could see four, five, or six of the huge boats threading their way in single file through the rocky islands of the Stockholm archipelago like elephants in a circus parade.

The *Estonia* was the second largest of the Baltic giants, a roll-on-roll-off ferry (a *ro-ro* in the parlance of the trade). Her overall length was 509.7 feet, just under one-tenth of a mile. She measured 79.38 feet in breadth. Her bridge stood ten decks above her keel, eight decks above the waterline. Fully loaded, she displaced 12,739 metric tons, and she measured 15,598 gross register tons. Despite her size and weight, she drew only 18.34 feet. For propulsion, she had four diesel engines, two to each of her counter-rotating propellers, delivering a total of 17,200 kilowatts of power at 600 rpm. She also had two bow thrusters for maneuvering when docking or leaving port and a stabilizer fin on either side, which could be deployed in rough weather and retracted in good weather to reduce drag. The stabilizers had been added in February 1994, a year after new management had decided to change the *Estonia*'s route from the sheltered path through the archipelago between Sweden and Finland to a direct run across the open Baltic to Tallinn.

At the *Estonia*'s very lowest level were tanks for fuel and other liquids, packed between the steel plates of the double bottom. Next came the so-called tank deck, or Deck 0, which contained a sauna and a swimming pool, the bow thrusters, and the lower half of the engine room. Deck 1, still below the waterline, housed the upper half of the engine room, the repair shop, and the engine control room in its rear half. The forward half was devoted to economy-class cabins for 358 persons. The quarters could not have been very pleasant for their occupants, who had to share communal toilets and showers located in the hallway, but the passengers were expected to spend most of their time on the upper decks shopping, drinking, and dining. Above Deck 1 was the main deck, or car deck, also known as Deck 2. Technically, the compartments below Deck 2 made up the watertight hull on which the ferry floated.

Deck 2, actually a double deck, was more than 4.5 feet above the waterline. Cars and vans were loaded first, onto a cleverly constructed false floor, Deck 3, which was elevated when all these relatively small vehicles had been accommodated, leaving room for the "big boys" down below on Deck 2. There was a great deal of long-distance truck traffic between the Baltic countries and Russia on one side of the water and the Scandinavian countries on the other. It was—and still is—widely believed that some of the truckers, helped by some of the ship's crew members, were engaged in smuggling.

Cars and trucks entered the big ferry by two doors at the stern, one for each side. They exited through an ingeniously designed bow that swung up like the visor of a knight's helmet and, in fact, was called a visor. Inside the visor was a loading ramp that was lowered for the cars and trucks to drive off the ship and was raised to a near-vertical position when the ship was ready to depart. The ramp rested against inboard supports when in the upright position. It was designed to withstand many tons of force from waves in case the bow visor failed. A locking arrangement allowed it to act as a barrier, keeping the motor vehicles from rolling overboard in case their parking brakes and tie-downs failed. Decks 2 and 3 were termed the *weathertight superstructure* of the ship.

Above the parking decks was the deck house with Deck 4, where the better passenger cabins started, continuing up on Decks 5 and 6, where the bars, shops, and other amenities were also located. The crew's quarters were on Decks 7 and 8, where the rolling and pitching of the ship were felt more strongly. The bridge stood in lonely splendor on Deck 9. Perched atop it were the radar and other navigation equipment.

The *Estonia* was well equipped for safe navigation. She had five radars, two gyrocompasses, two autopilots, an echo sounder for depth information, three global positioning satellite (GPS) receivers, and a navigation computer, plus other sophisticated equipment. Her fire protection system, however, was out of date, consisting only of smoke detectors and no automatic sprinklers. A security system with multiple TV cameras gave assurance against theft and other malfeasance on board. Below Deck 2, her floating ability was backed by thirteen watertight bulkheads (although there were twenty-two doors fitted in them) and fourteen watertight compartments or tanks.

The ship had been built in 1980 at the shipyard of Jos. I. Meyer in northwestern Germany for a big Finnish company and was named *Viking Sally*. (The choice of name was purposeful, for the shipping company was called Rederiaktiebolaget Sally.) After being shuffled among various subsidiaries, with a change of name each time, she was sold in 1993 to an Estonian firm called Estline, and her name was changed to *Estonia*. The officers and most of the crew were Estonian, and Estonian was the working language of the ship. Estonian is not a major language, and few, if any, of the Swedes, Germans,

Finns, and others among the passengers could understand it.[16] The captain and several of the other officers, however, could speak Russian since they had been trained at a Russian maritime academy. (It should be noted that, from 1940 to 1991, Estonia was part of the Soviet Union, which had annexed it, although the Germans occupied it during World War II.) The captain could also communicate in Swedish, Finnish, and English.

It was a gray, windy evening when the *Estonia* left Tallinn at dusk on September 27, 1994. Some of the passengers headed directly for the shops and bars, while others settled into their cabins, and some apprehensive souls took seasick pills. On this trip, the passengers were mostly Swedes and Estonians. Among them were fifty-six retired citizens on a group excursion, twenty-one teenagers from a Bible school, and most of the members of the municipal council of the Estonian city of Võru.

The weather became rougher as the *Estonia* proceeded westward out of the harbor, but the more determinedly upbeat among the passengers were enjoying themselves in the bar, the swimming pool, and the sauna. In the bar, the band was playing, and some of the passengers were dancing, ignoring the slight rolling of the ship. By 9 P.M., however, the waves were becoming pretty high—some reached 20 feet—which was too much for most of the merrymakers. Many decided to call it a night and went to their cabins; some became seasick. But no one was worried; the *Estonia* was a big, modern ship and bound to be safe.

By 11 P.M., the big ferry was approaching the halfway point of its route, an area that had always been a dangerous one for shipping. The winds typically blow from northwest to southeast or from southeast to northwest, which means that a ship sailing on an east-west course is pretty much broadside to the waves. And the relatively shallow water increases the height of the waves as they "feel" bottom. The ship began rolling heavily, but a nine-member Estonian dance troupe gamely persevered through their whole repertory. It was their first performance, and they wanted to make it a good one.

---

[16] Estonian is similar to Finnish but different enough that speakers of one language understand very little of the other.

After the show, one member of the troupe, a twenty-three-year-old man named Risto Ojassaar, relaxed in an upstairs bar with a beer. A lover of the sea, Ojassaar was entranced by the storm. As he testified later, he pressed his nose against a porthole and watched the waves, like walls of water, assault the ship. Some of the waves appeared to reach as high as Deck 8, he thought. "I had always loved sea storms," he later told an interviewer, "and I was sitting there thinking, 'Oh, look at that wave! This is great!'"

The ship was traveling at 15 knots despite the waves, risky and poor seamanship under such conditions.[17] According to one report, the *Estonia* was an hour behind schedule, and the captain was determined to make up the lost time.

Midnight came, and the storm continued. By 12:30 A.M., the rolling was so bad that the band could not continue playing. Many of the passengers were suffering from nausea. Twenty-five minutes later, many of the passengers and at least one crew member heard two loud bangs. No one knew what it was. (The official report issued after the accident stated that it was the bow visor breaking loose from its locks and hinges, though it remained attached to the ship by the hydraulic cylinders that raised and lowered it.)

The crewman, who was on the car deck, reported the sound to the bridge and checked the ramp. He saw nothing out of the ordinary and assumed there was no danger. In the Pub Admiral on Deck 5, glasses hanging above the bar and standing on the tables began to fall. An entertainer on the ferry's hotel staff was leading a karaoke competition and sing-along. Though it was scheduled to end at 1:00 A.M., the entertainer said the event would continue for another fifteen minutes since everyone appeared to be enjoying themselves.

At 1:05 A.M., the visor was dangling precariously. According to the official report, as it swung it struck the ramp and knocked it partway open. The sea began to enter the car deck through the gap. A crewman in the engine room, looking at the surveillance TV, saw water on the car deck. He thought it was rain water and turned on

[17] Two other ferries, the *Silja Europe* and the *Mariella*, both from Helsinki, were traveling at about the same speed, so perhaps risk taking was a regular practice on this route.

the pumps. Not considering the situation serious, he didn't notify the bridge. But the pumps could not keep up with the influx of seawater, and the engineer on duty went in person to the car deck. He found the water already up to his knees.

At 1:15 A.M., the bow visor fell off completely. On its way down, it crashed into the ship's hydrodynamic bulb, which extended from the bow below the waterline like an overdeveloped chin. Many passengers heard the noise, which survivors later compared to a giant sledgehammer striking the hull, making it reverberate like a bell.[18] Some of the people in the Pub Admiral said jokingly, "We've hit an iceberg!" One apprehensive fellow wanted to leave the bar, but his friends persuaded him to stay.

Another blow followed within a minute, and the ship began to roll, pitch, and yaw. A pronounced list to starboard developed as the flooding increased, soon reaching 30 degrees. At this point, a weak female voice came over the intercom system and said, "Häire, häire, laeval on häire!" Estonian for "Alarm! Alarm! There's an alarm on the ship!" Only the Estonians understood this message. It is believed that the woman, who was probably a staff member on duty at the information desk, was about to repeat her announcement in English when the P.A. system failed.

Some of the passengers and crew were already beginning to make their way up to the open areas of Deck 7, but as the list increased it became almost impossible to get to the stairs. Risto Ojassaar, reading a newspaper in his bunk, was thrown violently to the floor and landed on his neck. He looked up and saw the cabin door over his head. With his dancer's conditioning, he was able to grab the door frame and pull himself out. Others were not as fortunate. Some passengers found their doors blocked by baggage and furniture that had shifted. Others crowded into the stairways until they became so jammed that they were impassable. Some of the crew collected their wits sufficiently to try to organize an orderly evacuation, but because the ship was lying almost on its side they did not have much success.

---

[18] Other witnesses said that it sounded like a shot. It is normal for different witnesses to have different recollections, sometimes wildly so.

The manager of Ojassaar's troupe, who was in a nearby cabin and had sailed on the *Estonia* before, realized that they could not escape on the main stairway with the ship leaning so far over. She guided Ojassaar to another flight, and they managed to pull and heave each other up while the stairs rocked and lurched beneath their feet. At one of the landings, Ojassaar fell against a glass door to the outside. Looking out, he saw a swath of raging sea below him and realized for the first time what danger he was in. At last, the two Estonian dancers reached an outer deck. Before they had time to put on their life jackets, a huge wave washed them both overboard and separated them. The next time Ojassaar saw the woman who had saved his life was at her funeral.

The main stairway was pure chaos. Injured passengers competed for passage with drunks and age-slowed senior citizens. Everyone who could reach it was grabbing for the handrail, which came loose under the load. Only the fittest could claw their way up. The elderly and the children never had a chance.

In the barroom, a loud scream was heard as the heavy bar counter came loose and crushed the barmaid against a wall. In other parts of the ship, slot machines broke loose from their fastenings and fell on helpless passengers. Other passengers saw them but could not help them since they dared not release their grips on whatever had not yet broken loose.

The second officer sent out a *Mayday* call at 1:22 A.M. It was picked up by another ferry, the *Mariella,* which responded immediately. The officer, however, did not communicate his problem or his position for another two minutes since he was trying to contact more vessels. At 1:24:31, he said, "This is *Estonia. Silja Europa* [another ferry], *Estonia.*" *Silja Europa* identified herself, and the *Estonia*'s officer continued. "Good morning," he said, "do you speak Finnish?"

"Yes, I speak Finnish," came the reply.

"Yes, we have a bad problem here now, a bad list to the right side. I believe it was 20, 30 degrees. Could you come to our assistance?"

The other ferries asked for *Estonia*'s position. But a power failure on the *Estonia* held the transmission up for agonizing minutes. Finally, the *Estonia*'s first officer was able to radio "59 latitude and 22 minutes." Asked for the longitude, he said, "21.40 East."

Silja Europa replied, "21.40 East, okay."

The *Estonia*: "Really bad, it looks really bad here now."

Silja Europa: "Yes, looks bad. We are on our way, and it was 21.40."

Estonia: "...you said..." That was her last message to the world.

The *Estonia* began to list so severely that those passengers who could climbed up on the side of the hull, wet and slippery as it was. Inside, some passengers were in shock, unable to move. Chairs and tables tumbled down like missiles, injuring many. Survivors recalled heartrending scenes. In the Café Neptunus, on Deck 5, an elderly woman and her adult son painfully made their way up the floor, which was almost vertical, pulling themselves up by the pedestals of tables bolted to the floor, as if they were climbing a nightmarish set of monkey bars. The son reached the doorway and turned around; his mother, clinging to a table pedestal, said she didn't have the strength to go on. He should save himself, she told him. He managed to do so, but she drowned. Another survivor tried to lead his parents and his girlfriend across the foyer; he struggled with the other panicked passengers to make a way across. He looked back and saw that his loved ones were still on the main stairway, clinging desperately to a section of the handrail, trapped by the crowd. They screamed to him that he should save himself, and he left them behind. One wonders how he felt the next day.

Before the ship heeled over completely, the captain turned her to port so as to face into the wind and the waves—a recommended safety maneuver—but it was too late. One by one the engines cut out, flooded. The *Estonia* began to drift eastward. Later some of the official investigators thought the captain might have been trying to turn the ship back to Tallinn, but his true intention will never be known.

From the moment the *Estonia* heeled over beyond recovery, the passengers had an estimated fifteen to twenty minutes to save themselves. Fewer than 300, including crew, managed to get out. Seven hundred and fifty were trapped inside; the people on deck heard their screams for help. Out on the deck, the crew tried to launch the lifeboats, but the davits were rusted shut. They pushed life rafts overboard, but many drifted away, and others were flipped upside down by the waves. The people on the upturned hull had the choice of sliding down the hull into the deadly cold water or sliding the other way and smashing against the deck. The life jackets—there were more than enough to go around—dated from the original *Viking Sally*; they were old and poorly designed. Many passengers couldn't fasten the crotch straps, and their life jackets slid off. Some jackets were too large or too small; some were tied together in bundles of three, and the passengers could not untie them in the dark and the cold. Some passengers reverted to utter selfishness and tried to tear the life jackets off the nearest fellow passengers, even though plenty were lying on the deck. Meanwhile, waves kept washing people, and life jackets, into the sea. Thus, a number of people found themselves thrashing about in the water with no flotation device to help them despite being surrounded by floating, empty life jackets.

So unprepared were the passengers that many of them tried to flee in their pajamas or nightgowns. One survivor saw a pair of young women standing, paralyzed, in the corridor, dressed only in their panties. He did not see them again. Others did not have time to put on their shoes. Some were entirely naked.

The hard plastic sheathing of the deck broke up and came loose, slid about, and prevented people from moving. Then it slid down over the open aft doors, blocking them. The *Estonia* lay at a 45-degree tilt for a few minutes, then abruptly heeled over to 80 degrees. Passengers on the outside slid and jumped into the sea, while others did not need to make even that effort—the waves washed them overboard.

The moon had come out and gave enough light so that the people struggling in the water could see their ship's white hull against the dark sea. When the list had increased to about 135 degrees, several lifeboats broke loose from their supports and fell against the ship, which damaged them severely. A number of the people still on board fell into a tangle of wires and the ship's cranes and loose lifeboats. The water was now full of life rafts, which had also come loose, empty life jackets, and terrified human beings.

There were at least ten life rafts tossing about in the frigid water. The problem was getting into them: many were upside down, and people who managed to crawl onto them had only the most precarious of holds. Waves kept washing them off, and they did not always make it back on. A man who had been washed off one raft swam for about an hour before making his way to Life Raft B. Several life rafts with people in them had drifted by him, but no one had the remaining strength or coordination to throw him a rope. Wearing four life jackets, he called for help. Another man on Life Raft B heard him but could not find him in the dark. Finally, the exhausted swimmer was pulled aboard. At first, the two men sat on top of the raft's canopy, which was intended to shelter them from wind and water. It could not as long as they were sitting on it. They eventually managed to crawl underneath the canopy, raise it, and close it. They took turns bailing water out of the raft's bottom with a small plastic scoop that they found. After a while, they discovered a plastic bag that held a flashlight and another bailing scoop. Their hands were too stiff with cold to open it, so one of them tried to tear it open with his teeth. He gave up after several of his teeth were torn out.

Life Raft I held many people. Originally, about fifty passengers and crew had jumped into it while it was still on the deck of the *Estonia*. As it slid off the deck, most of them fell off; some managed to get back on. But even on the raft their situation was perilous. Waves kept crashing over them and flooding the raft. A few tried to bail but made little headway. After about forty-five minutes, they managed to get the canopy up and closed, which made things a little better. They could hear people outside crying for help, but they could not see enough in the dark and the waves to assist them. One of the crew members on this raft had a flashlight. The other occupants of the raft wanted him to signal for help with it, but he refused. Some of the more energetic people struggled with him and eventually managed to pry his fingers loose from the light.

Inside the raft, people tried to lie close together for warmth, but they sloshed back and forth in the water that filled it. Someone found signal rockets and fired them off, but there was no one to rescue them. At last, after several hours, daybreak came, and the exhausted survivors saw another ferry, the *Isabella*, eastward bound from Stockholm to Helsinki.

The *Isabella* put out a life raft with three crew in survival suits. These men shouted to the people in Life Raft I to come over to the *Isabella*'s raft, which could be hoisted right on board by a rope on top. The *Estonia* survivors were too weak to respond at first, even as the waves kept slapping their raft against the *Isabella*'s hull. Finally, they came across, a few at a time. By now the *Isabella*'s raft, too, was filling with water and becoming very heavy. The *Isabella*'s winch could not lift it. The three crewmen in survival suits jumped into the water to lighten the load, and after about ten tries the raft began to rise. But the bottom split, and all but one of the occupants fell through it back into the sea. Several came up underneath the raft; a few, with no strength remaining, disappeared under the water. Eventually, someone on the *Isabella* thought of opening a port in the hull and deploying an evacuation slide, which the crew aboard the ferry helped the survivors scale. Sixteen were rescued in this fashion, and one was picked up by a helicopter.

Life Raft K was floating upside down. Two women in their twenties had somehow managed to get on top of it. Then a male Swedish passenger, who had been washed from the *Estonia* by a wave, climbed on. They were joined by an Estonian repairman from the crew, who had to swim beside the raft for ten minutes before the others were able to haul him up. Waves washed them off four or five times, but each time one of them managed to keep a grip on the raft, and they all got back on. They lay close together and hugged and massaged each other to keep warm. Although it sounds like an erotic encounter, it was anything but. The men proposed that they all go back into the water and flip the raft right side up, but the women said they were too afraid to go back in the water again. Soon after, another wave broke over the raft, and the two women slid off and drifted away in the darkness. One of them was moaning as she was carried away. At least twice more waves washed the two men overboard. The last time, the Swede got caught in a rope, but the repairman was able to get back aboard. He tried to help the Swede up, but they were both too weak. The repairman took the Swede's hands and kept his head above water until they were rescued by a helicopter about 7 A.M.

Similar stories were heard from the survivors on the other rafts and in the few lifeboats that floated, mostly upside down. In one raft, the people lay in vomit, both their own and others'. A number of people died of exhaustion and hypothermia in the rafts.

The investigating committee found grave deficiencies in the survival equipment: life jackets that did not fit or would not stay on or had missing straps; aluminized plastic garments, meant to retain body warmth, that ripped when people tried to put them on; knives that would not cut entangling ropes; life rafts whose sides were too high to climb over, canopies that could not be closed, and bailing scoops that were so small and ineffectual that some passengers bailed water with their shoes instead.

Of more than 200 people who had made it onto the deck, 138 were rescued. (One died later in a Swedish hospital.) Ninety-five bodies were recovered and identified. Seven hundred and fifty were listed as missing; most of them were trapped inside the hull of the *Estonia*, but some had just vanished. Among the dead were 461 Swedes, 237 Estonians, 13 Latvians, 10 Russians, 9 Finns, 6 Norwegians, 5 Danes, 4 Germans, 3 Lithuanians, 2 Moroccans, and 1 each from Belarus, Canada, France, the Netherlands, Nigeria, Ukraine, and the United Kingdom.

The sinking itself was witnessed by Risto Ojassaar and others. At 1:50 A.M., the gleaming white ferry sank stern-first amid a trail of thousands of bubbles.[19] Many witnesses said later that they noticed the bow visor was missing (others said that it was not). The witnesses also saw people clinging to the outside railings on the ship, but not letting go when she went down. The captain drowned at his post on the bridge, but not alone—everyone else on the bridge died with him. The *Estonia* came to rest on a bed of soft clay, lying on her starboard side with a 120-degree tilt past vertical, at a depth of 243 to 279 feet.

The day after the accident, the governments of Sweden (Swedish companies were the majority owners of the ferry line), Estonia (flag country of the ship), and Finland (in whose search area the ship had sunk) formed a Joint Accident Investigation Commission (JAIC) composed of three representatives from each country. In the opinion of

---

[19] This is the time at which she allegedly disappeared from the Finnish radar that was monitoring the area. The Finnish record of the event has disappeared. By 1:30 A.M. (one of the surviving officers checked his wristwatch), the *Estonia*'s stern was under water, and her bow was rising. She slid under at a 45-degree angle, belly-up.

many, the JAIC busied itself immediately with a grand whitewash job—but more about this later in the chapter.

The investigation began, logically enough, with locating the wreck, which was accomplished by a hydrographic research vessel belonging to Finland. The searchers had to contend with heavy weather, but with side-scan sonar and a multibeam echo sounding system they found the wreck on September 30. An ROV inspection began on October 2—again delayed by bad weather. JAIC met on October 3 and 4 at the Finnish port of Turku and decided that more video images were needed to ascertain the general condition of the *Estonia* and whether her bow visor had in fact fallen off. It should be mentioned that the survivors were interrogated multiple times by the Finnish police, Estonian police, JAIC members, and, in some cases, Estonian security police. The crew members were questioned most intensively, and some news media believed they were threatened to keep them from speaking up about potential concerns regarding the ship's safety.

According to the JAIC report, the visor was located by sonar on October 18, about 1 nautical mile east of the wreck. Videotapes from an ROV confirmed that the massive steel object was the visor. On November 18, a Swedish minesweeper and a Finnish icebreaker recovered the visor, which weighed 55 tons, and brought it ashore to Turku for investigation. Subsequently, the Swedish government ordered a diving survey of the wreck, which was done by a commercial diving firm called Rockwater between December 2 and 5. The announced purpose of the survey was to determine the feasibility of lifting the wreck and recovering the bodies of the victims.

The observations made by the divers were recorded in written reports, supplemented by the ROV's videotapes. In addition, the divers cut away and recovered a number of pieces from the damaged bow: the housing for one deck-hinge bushing; all three attachment lugs from the visor's bottom locking device; the locking bolt for the bottom lock; a failed hinge lug from the inner ramp hinge (port side); and two steel spacer rings from the failed ramp hinge. They also brought back one emergency position indicating radio beacon (EPIRB) storage case, one GPS receiver, a portable lifeboat radio set, and a ship's bell.

The lugs from the visor's bottom lock originally had holes in them for the locking bolt to pass through. The divers found them all broken, with no holding power. The hinges of the ramp showed marks of heavy pounding. The rubber seal around the opening in the bow was broken in places, and there was other damage. As for the visor itself, it bore a massive dent slightly to the right of its center line, a continuation of a matching dent on the stem. The white paint of the visor was marked by blue paint it had apparently picked up by banging against the lower portion of the ship. The visor's rusted bottom, showing patches of various colors of paint, had been bashed and distorted until it resembled a piece of "non-representational art."[20] The list of damage to the visor and bow continues, item by item, for most of ten pages in JAIC's final report.

These evaluations were only the first steps in the investigation. The recovered pieces of metal were examined microscopically for signs of stress. The experts found that the solid parts, such as the lugs and the hinge beam, had failed because of ductile stress (from pulling or pushing or bending to the point of overload) rather than from metal fatigue. Some of the welded joints had been attacked by slow interior corrosion, and some showed bad fatigue cracks. According to the report, the welding seemed to have been so unevenly carried out that measuring the welds was "very tedious." Engineers also made complex calculations of stresses on various other parts. The paint had a story to tell, too. Analysis of the different layers indicated that the bottom lock was old and dated back "at least to the early history of the ship." This observation implied that the German shipyard had used a secondhand component, an allegation that the shipyard strongly denied. According to the Germans, the locks and hinges on the visor and ramp failed because they had never been maintained properly, which appears to be highly probable.

The Swedes went to the trouble of constructing a scale model of the *Estonia*, at a scale of 1 to 35. The model was self-propelled, and the bow visor was fitted with a six-way motion sensor. The model was tested in a towing tank to see how the bow visor behaved with

[20] A full-color photo of this damage can be seen on p. 122 of JAIC's *Final Report on* ... MV Estonia.

the seas coming from head-on and in a maritime dynamics tank to simulate waves coming from the side or at an angle. The investigators studied the effects of the waves by computer simulation and also created computer models of the sinking to see how various scenarios played out.

The conduct of the sinking ferry's evacuation was evaluated, as was the weather forecast and the loading of vehicles onto the car deck. Surviving crew members told the commission that the trucks had been directed so that there was more weight on the ship's starboard side than on the port side. To compensate for this imbalance, the port trim tank was pumped full. Even so, the *Estonia* left the pier with a slight list to starboard. Why the unbalanced parking arrangement was allowed will probably never be known. The radio communications and the response time of the rescue forces were also analyzed. In short, there was virtually no detail that the report did not cover.

The report was issued in 1997 under the title *Final Report on the Capsizing on 28 September 1994 in the Baltic Sea of the Ro-Ro Passenger Vessel* MV Estonia. It is a handsome production, with a beautiful layout, color photographs, and diagrams. It concluded that the bow visor's attachments were not designed for the severe conditions that the ship encountered on the fatal night, and, as a result of the failure of the visor, large amounts of water entered the superstructure, causing "loss of stability and subsequent flooding of the accommodation decks." The crew came in for some mild reprimands for laxity in observing safety regulations and slowness in responding to the emergency. The bridge officers (no one person was singled out for criticism) were blamed for not reducing the vessel's speed and not ordering an investigation of the bow area immediately upon getting reports of the metallic sounds. On the other hand, the report pointed out that from the bridge they could not have seen whether the visor was closed or open. The personnel on the bridge (now conveniently dead) were posthumously criticized for being late in ordering the evacuation. The Finnish marine rescue centers at Turku and Helsinki were reprimanded politely for being slow in sending out the helicopters.

The report concluded with several recommendations: better design of bow visors for ro-ro passenger ferries; better gathering and dissemination of information on accidents; guidelines for stormy conditions to be issued to all passenger ferries; reevaluation of evacuation procedures with an eye to streamlining them; new and more effective designs for lifesaving equipment; and regular practice on distress calls and responses for deck officers and radio operators at rescue centers. No one could quarrel with these recommendations, yet the report was attacked as soon as it came out. In fact, groups and individuals had begun attacking the JAIC and its work almost as soon as the committee was formed.

All of the JAIC's opponents agreed on one point: the committee and its report were essentially a cover-up operation. The press and the Internet seethed with theories and rumors about the real cause of the *Estonia*'s sinking. One story claimed that two undocumented Russian trucks had been on board, loaded with high-tech military devices and radioactive materials. These vehicles were destined either for the CIA or for Israel (the anti-Semites had a field day with the rumor), with Sweden as a transshipment point. In either case, the goods were allegedly bought from corrupt Russian military officers. Loyal Russian officers tried in vain to stop the shipment. Instead, they either set off one or more bombs or had the bow visor raised and the ramp lowered in the middle of the Baltic and drove or pushed the trucks off the ferry—anything to keep this dangerous cargo out of the hands of the enemy. Waves rushed in through the opening and sank the *Estonia*—the perfect cover-up. However, the likelihood of someone opening the bow visor and driving—to say nothing of pushing—a heavy truck out the bow while waves 15 feet high were slamming into the opening is not great.

Others disagreed and said the sinking was a terrorist act. Some thought smugglers with a massive load of heroin, afraid that the authorities had been tipped off, blew a hole in the hull to avoid being caught with the goods. Still others theorized that the Estonian branch of the mob had been extorting protection money from the Estonian government. When the government stopped paying, the mob arranged to have one or more time bombs go off on the *Estonia* as a lesson to those who wouldn't pay. More sober critics insisted that the JAIC was covering up for massive incompetence and corruption by officials of the three countries involved.

One of the most persistent critics was a Swedish naval designer and maritime safety expert named Anders Björkman, who published several books and pamphlets that dealt with the circumstances of the sinking. In page after page of close reasoning, spiced with occasional sarcasm, he maintained that the official theory for the *Estonia*'s sinking, the flooding of the superstructure, was nonsense. He pointed out that such flooding would make the ship turn upside down; buoyed by the air in the undamaged bottom, it would float indefinitely (as was the case with the Polish ferry *Jan Heweliusz*, which capsized in 1993). To make the ship sink, he said, the lowest decks had to be filled with water, which could only have come through a leak, one perhaps caused by rust or defective repairs to the plates of the hull or by something else. The JAIC report said nothing about an underwater leak, although passengers on the cheap-fare Deck 1 had found water coming in there through open watertight doors that were supposed to be kept closed. But nothing would have leaked from the car deck to the decks below because the car deck itself was watertight; the only passageway to the lower decks was through the stairwells, which were in the central housing of the ship. And those stairwells would have been unaffected because the incoming water would have collected at one side of the tilting car deck and superstructure.

Björkman found many other holes in the JAIC report, such as contradictory testimony and changed stories from surviving crew members. There was also the matter of where the bow visor fell off. The Finnish search ship reported it east of the main wreck, whereas it was finally located 1 nautical mile west of the wreck. The commission blamed this discrepancy on Finnish incompetence. However, the captain of the Finnish ship said later in confidence that he had been pressured to give a false position.

There was a belief among some experts that the starboard stabilizer fin had broken off and let water into the hull. Björkman noted that the stabilizer fins had been installed in January 1994, a hasty job that was never approved by the Estonian authorities. But he did not believe that the fin had come off. His personal belief was that there had been a severe leak caused by an explosion—but not a bomb—perhaps in a sewage tank, the drainage system, or a collection tank for drainage from the car deck. Such accidents at sea are frequent, he said. He theorized that some hazardous material leaked from one or more trucks, drained into a collecting tank, and was ignited by an accidental spark. The resulting explosion would have created a long gash in the ship's side low enough to flood some of the compartments on the lower decks.

Björkman poured scorn on the story that the ramp had opened up to admit a flood of seawater into the superstructure and the car deck. He noted that none of the survivors had seen it open as the ship sank and that the government's divers could not get into the hull through the bow opening because the ramp was closed and blocked their way. Was it likely that the ramp had come open and then closed itself again? He postulated that the ramp was probably a bit out of line and leaky, but an alert crew would have been able to make it tight by stuffing the cracks with whatever was handy and securing it with a rope so that it could not be ripped open by the visor.

One of Björkman's strongest points was that the *Estonia* had never been intended for travel on the open sea. The ferry was originally designed to travel through the sheltered waters between Stockholm and the tourist haven of Mariehamn, in the Åland Islands.[21] Later it was sent on longer routes, but always within 20 miles of land. When the ship was sold to Estonia, it should have been reinforced for the rough water of the open Baltic, especially in the bow. It should also have been given proper lifesaving equipment, and most of the watertight doors down in the hull should have been welded tight. The Estonians, apparently wishing to economize, never made these improvements. Instead, they took a gamble on a ship that was not seaworthy for her new route.

Björkman also criticized the makeup of the JAIC and its modus operandi. By international law, Estonia, as the flag state, should have directed the investigation. Instead, Sweden took over. There were also many changes in the composition of the commission. Individuals

---

[21] The Åland Islands lie between Sweden and Finland. Although the islands belong to Finland, the population is Swedish-speaking, a relic of the days when Sweden ruled Finland.

retired (or, as Björkman said, perhaps were kicked off) for no apparent reason. At one point, the head of the Finnish delegation was an official from the Finnish Security Police, not a maritime expert. Toward the end of the investigation, one of the Swedish delegates was a psychologist who knew nothing at all about ships or the sea—supposedly, he was needed to evaluate the testimony of the witnesses.

The actions of the commission did not inspire confidence. It announced its conclusion that the *Estonia* had sunk because of the failure of the visor and ramp only seven days after the accident, before the evidence was gathered and interpreted. It suppressed reports by two of its members, classifying as secret their belief that water on the car deck would have made the ship turn upside down and float rather than sink. The final report, delayed for three years, cannot even be debated publicly in Estonia.

In a very peculiar action, the diving/recovery team threw the central bolt of the bottom visor lock back in the ocean. The reason given was that the bolt was large and heavy, and the helicopter did not have room for it. However, the commission's team managed to find a place for the ship's bell, which was much larger and heavier. The bolt, through its wear marks, could have provided valuable clues to the visor's state of repair. And it was Björkman who pointed out that, during the official Swedish diving/ROV expedition in December 1994, divers were sent down in the middle of the night to recover baggage from certain cabins, leaving the bodies behind. This fact was not revealed to the public.

Although the families of the deceased and a number of organizations, including the powerful International Transport Workers' Federation (ITF), asked for an independent investigation, the JAIC refused. Sweden, Estonia, and Finland hastily confected a treaty forbidding any diving on the wreck site. This treaty was binding only on citizens of the three nations, but it put a severe crimp in investigations and hinted that they had something to conceal. Sweden went so far as to declare the wreck a grave site, sacred to the dead, who must not be disturbed—this in spite of repeated demands by the families that the bodies be recovered. Then the Swedish government announced that, to keep the site from being profaned by intruders, it would be covered in concrete. Concrete is expensive, so the Swedish government decided to first bury the wreck under load upon load of sand and top the layers with concrete. Some sand was dumped, but the project was quietly abandoned.

Björkman had only contempt for the various conspiracy theories, pointing out, for example, that, if the Russians had wanted to stop the theft of classified weapons, they had only to call the Russian embassy in Sweden. The embassy would have notified the Swedish police, who in turn would have intercepted the stolen shipment on the Swedish side. Instead, he accused the JAIC itself of conspiracy in his 1999 publication *New Facts about the* Estonia, translated from the Swedish:

> I have written to various Swedish authorities and never gotten a proper answer or clear information. Official Sweden protects its incompetent civil servants in every way and at the same time [protects] the official, incompetent, and corrupt Estonia, which they have taken under their wing. Whether it is the corruption in the old Soviet Union that has spread to Sweden, or the old, calcified tradition of Swedish officialdom to never admit a mistake, which is the reason for the concealment I do not know.... But international safety at sea has been hit hard by Estonia's, Finland's, and Sweden's moral collapse and lies about why the *Estonia* sank. The work of international safety at sea cannot be founded on lies.... Someone should really stand up and say that the emperor has no clothes.

In 2000, Gregg Bemis, an American businessman with long experience in underwater search-and-salvage, and Jutta Rabe, a German journalist who had followed the *Estonia* story since 1994, joined forces in a privately financed expedition to the wreck. They used a former German oceanographic research vessel, *One Eagle*, which Ms. Rabe had leased, with a towed sonar mapping system and ROV supplied by Bemis. The dive was planned for late August, a quiet time on the Baltic.

One Eagle set sail on August 19 from the German port of Cuxhaven, which is located on the northwest coast of Germany, on the far side of the Danish peninsula. To avoid the long trip around Denmark, it sailed through the Kiel Canal. On board were Bemis and Rabe; four other Americans, who handled the sonar and ROV; a Swiss cameraman and his German woman assistant; two Germans from the major German magazine *Der Spiegel*, one a reporter and the other a photojournalist; and seven Germans and two Czechs, who made up the diving team. Bemis and Rabe also had a contract with *Der Spiegel*. The expedition reached the wreck site on August 22, only to find two Coast Guard ships from Sweden and Finland patrolling the area to chase off the nosy intruders. A number of other ships, carrying TV crews eager to see what was going to happen, crowded around.

The Coast Guard captains hailed *One Eagle* and asked to come aboard. They warned Bemis and Rabe that they would be arrested if they violated Swedish law and the tripartite treaty by diving on the wreck. Bemis replied that the site lay in international waters and that none of the personnel on board was a citizen of the treaty nations, so the treaty did not apply to them. The Coast Guarders were forced to agree. The Finnish captain was more concerned about the divers' safety, but after interrogating them he decided that they knew what they were doing and gave them permission to dive after extracting promises of good behavior.

The first step in the expedition team's plan was to take sonar images of the wreck. It was not to be, for the towfish's cable got fouled on the propeller of the ship and was cut in two. The expensive towfish sank to the bottom. Over the side went the ROV, which passed over a debris field and eight or ten bodies. Some still wore bits of clothing; others were just skeletons. Eventually, the sonar fish was found and its position noted. The ROV was pulled up and fitted with a hook to recover the towfish. The hook broke under the stress of the lifting attempt. The team marked the position of the towfish again and pulled up the ROV so that a new hook could be welded on.

By now the first team of divers was ready to go. The ROV led the way and found the stern of the wreck with no trouble. Unfortunately, strong winds made *One Eagle* difficult to maneuver, and visibility in the water was only 15 feet. The divers used the ROV's tether as a guideline, but strong currents shifted the ROV away from the wreck;

in the mud-laden water, the divers could not find it. The next day, conditions were too rough for diving. Instead, the ROV was sent down again to find the towfish, but it had disappeared. The search continued for several days, but the towfish was never found. Bemis suspects that the Swedish Coast Guard had picked it up under the cover of night.

Eventually, the stern and bow were found and marked with shot lines, and the divers began their real inspection. They found the propellers and the stern car ramp undamaged. There was no debris from the vessel in this area, but the divers could not document any of their observations since their handheld video camera malfunctioned.

On August 25, the ROV's tether got caught in the mother ship's port bow thruster and was cut off. The tether was cautiously unwound from the propeller, and the ROV was hauled in, but during the brief moment when it was cut by the thruster propeller a short-circuit damaged some important components in the control console. The expedition leaders, realizing that the American company that had manufactured the ROV ensemble could not possibly deliver replacement parts within the needed time, leased a small ROV system from a German company. The ROV was to be delivered to the nearest port, which happened to be Gdynia, in Poland. There a German boat picked it up and delivered it to *One Eagle*, but the wait cost the expedition three days of ROV downtime. The divers, however, were able to go down to the wreck and take pictures each day.

Two divers were sent forward along the *Estonia*'s port side to film the side of the hull and the port stabilizer fin. They did not have enough bottom time left to reach the stabilizer, but they did get some images of the hull. One of the divers found a hole cut in the hull that had been left uncovered. It would have been made by the official Swedish diving expedition in 1994 or perhaps by a subsequent illicit expedition.

Groups of two or three divers were sent to swim in formation along the starboard side to inspect it for holes or other signs of damage. They were able to cover the entire starboard side down to the mud line. They found the right stabilizer fin in place and undamaged (so much for the theory that it had broken off, causing a leak). The divers also found many scratch marks along the hull that had been made by ROVs and hoses. The marks were not very old, since they were not yet covered by sediment. Who had made them? The divers also found a large pile of sand between 3 and 4 feet high along the forward third of the wreck, the remains of what had been dumped by the Swedish government.

When the divers filmed the bow, they found a great deal of man-made damage whose origin could not be decisively determined. They also attached a device for measuring radioactivity to the underside of the ramp and cut off two small pieces of metal from the bow visor's bulkhead, of which the lock was a part. These pieces were later subdivided and sent to two laboratories in Germany and one in Texas to be tested for evidence of explosion.

Both Bemis and Rabe were convinced that the *Estonia* had been holed and sunk by explosives. The three labs to which they had sent their samples confirmed that the metal showed "deformation twinning": that is, the crystal structure had been altered by a tremendous mechanical stress—from high explosives, for example—which showed up under the microscope as a kind of streaking. But *Der Spiegel* engaged a fourth laboratory, this one operated by the German state, to do still another analysis. This lab tested an unexploded piece of the same type of steel from the original shipyard and found that its grain pattern matched closely the sample from the ship. Then the lab applied explosive to another piece of steel and found a much more marked pattern of disturbance. The answer was found in a practice of the shipyard: it blasted the bare metal parts with steel shot, a process called *shot-peening,* to get rid of rust before painting them; this process produced the explosion-like effect. Bemis's comment on the findings is that the state laboratory lacked experience in performing this type of test and didn't have the proper equipment. Furthermore, he says the effects of shot-peening extend only about 1 millimeter below the surface of the treated metal, while deformation twinning goes deeper. The explosives, Bemis believes, were used by the Swedish Navy to blast the bow visor free so that it could be moved to a bogus location in order to bolster the official theory that the *Estonia* sank because the visor fell off.

Meanwhile, the replacement ROV arrived at the wreck site late at night on August 28. By then, the supply of diving gases was almost exhausted, and so were the divers. But they went down one final time on August 29, though they did not find much that was new. They did note that the wreck had leaned over about 30 degrees more since the date of the tragedy and was resting at 150 to 160 degrees instead of the 120 to 130 degrees originally reported.

On the very last day of the expedition, the ROV did some filming around the bow. The viewing screen in its control van aboard the mother ship showed what appeared to be a hole in the hull, though it was mostly hidden by the sand. Others who saw the images later insisted that the hole was really a shadow cast by the ROV's floodlight. Until someone can investigate the wreck of the *Estonia* again, its identity will remain a mystery.

For reasons unexplained, *Der Spiegel* backed out of its deal with the two organizers of the expedition. Bemis insists, that in spite of bad weather and equipment problems, he succeeded in getting plenty of photographic, material, and other evidence.

Rabe, thwarted in making a documentary for German TV, fictionalized the story as a movie instead. Its main characters were a truth-seeking female German journalist; a Swede who had lost his son in the wreck, for love interest; and another Swede who had lost his wife and was intent on recovering her body so that they could be buried together, for tragic interest. It was not a very tasteful tribute to the dead of the *Estonia.*

In April 2003, a retired Swedish admiral published what he said was a hitherto ignored set of engineering drawings of the *Estonia.* They showed six giant ventilation shafts on each side of the ship, which purportedly supplied fresh air to the engine room and the other spaces below the automobile deck. Their intakes were placed just below Deck 4. In this position, they could take in water when the ship heeled over as little as 38 degrees. Thus, the admiral said, the great waves must have poured down the ventilation shafts and flooded the ship. However, the drawings proved to be a clumsy forgery of unknown origin.

Was the sinking of the *Estonia* an unexpected marine tragedy or a deliberate plot carried off by parties that remain unknown? The mystery persists—as does the suspicion.

* * *

When the *Edmund Fitzgerald* went down in 1975, the death toll was 29; nothing close to the 852 lives that were lost in the *Estonia* tragedy. But the name of this huge ore carrier still stirs powerful emotions in the Great Lakes shipping community.

The *Edmund Fitzgerald* was built in 1958 at the Great Lakes Engineering Works at River Rouge, Michigan, on the narrow Detroit River, which connects Lake Erie to Lake St. Clair. From there, a ship can sail via the St. Clair River up through Lake Huron and, threading its way through several islands and the famed Soo Canal, to Lake Superior. It was on Lake Superior, largest of the Great Lakes, that the *Fitzgerald* spent her working life and met her end.

The launching of the *Edmund Fitzgerald* on June 9, 1958, was attended by much ballyhoo. Ten thousand people are said to have turned out to see the event. The giant ore carrier was rather bombastically proclaimed "the pride of the American Flag." She was named for the president of the Northwestern Mutual Life Insurance Company, which owned her. Bombast aside, there was reason to be impressed by the new ship. She was 729 feet long (almost the length of a 72-story skyscraper laid on its side) and 75 feet wide, giving her the typical long, slim shape of a bulk carrier—if the word *slim* can be applied to anything of that size. She could carry over 27,000 tons of iron-ore pellets; on her final trip, the load was recorded as 26,116 long tons (29,250 regular tons). Loaded for her last run, she drew 29 feet 2 inches forward and 29 feet 6 inches aft. A 7,500-horsepower steam turbine turned her giant 19.5-foot propeller, pushing her at a cruising speed of 16 miles an hour or more. During her career, the *Fitzgerald* underwent a few makeovers. In 1969, she acquired a diesel-powered bow thruster; in the winter of 1971–72, her fuel was changed from coal to oil; and she twice had surgery to correct cracks in her keelsons.[22] Welding was the method used for the second keelson repair in 1973–74.

[22] A keelson is a girder that runs along the top of the keel to reinforce it. A vessel like the *Fitzgerald*, with its bathtub-like cross section, would have a keel and keelson on either side. Considering the loads that lakers carried, and their tendency to grate frequently against the bottom when coming in to the loading

The master of the *Edmund Fitzgerald*, Captain Ernest McSorley, was a veteran mariner nearing retirement. He had worked for the operator of the ship, the Columbia Transportation Division of the Oglebay-Norton Company, since 1938 and had been a captain since 1951. He had been the master of the *Fitzgerald* since 1972. McSorley seems to have been generally respected by the men who worked under him, although one source said that he liked to push ahead through any kind of weather to keep on schedule and was lax about seeing that repairs to the ship were made. Apparently, the management liked McSorley because he kept to schedule and knew ways of shaving costs.

November is the month when storms on Lake Superior are at their most severe. But it was a pleasant, sunny afternoon on November 9, 1975, when the *Edmund Fitzgerald*, her cargo holds full and carefully trimmed to avoid stressing the hull, left the loading pier at Superior, Wisconsin, at the western tip of Lake Superior. Her destination was Detroit, and she had to traverse all of Lake Superior and Lake Huron to get there. Storm warnings had been issued by the National Weather Service, but McSorley apparently thought he could beat the storm on this voyage. First the National Weather Service predicted that the storm, which had originated over the Oklahoma Panhandle, would pass south of Lake Superior by 7 P.M. on November 10. But the service changed its mind six hours after issuing the original forecast; as of 7 P.M. on November 9, it issued gale warnings for all of Lake Superior. The warning meant that all ships on the lake would face winds of 34 to 47 knots, with proportionally high waves. High waves can endanger even the most stoutly built ships, and lakers were typically constructed with only half the longitudinal strength of seagoing ships. This design made them vulnerable to hogging (where the ship's middle is lifted up on a wave while the ends are unsupported) and sagging (where the ship's ends are lifted up and the middle is unsupported), both of which seafarers try to avoid.

By 1 A.M. on November 10, the *Fitzgerald* had reached a position about 20 miles south of Isle Royale, Michigan, and was reporting

---

and unloading piers, it is not surprising that the keelsons developed metal fatigue and cracked.

easterly winds of 52 knots and waves 10 feet high. As the ship's freeboard was 11 feet 6 inches, the waves came uncomfortably close to the deck. The Coast Guard, probably under political pressure from shipowners who wanted to carry bigger loads, had decreased the minimum freeboard requirement from 13 feet 3/4 inches in 1971 to 11 feet 6 inches in 1973. As part of the deal, all vents were to extend at least 30 inches above the weather deck, which on a bulk carrier is the top of the main deck. But as a further concession, the *Fitzgerald* was allowed to have ballast tank vents only 18 inches high because the taller size interfered with cargo handling. These shorter vents, which had mushroom caps like those on the *Derbyshire*, were more vulnerable to being overtopped by waves.

Isle Royale is only a fraction of the distance from the Wisconsin port town where the *Fitzgerald* had taken on her cargo to the Soo Locks. The big laker had many miles to go before reaching the locks, which were the halfway point to her final destination. By 7 A.M., she had struggled as far as the longitude of Copper Harbor, on Michigan's upper peninsula. The wind had moderated to 35 knots, but the waves were still 10 feet high. The *Fitzgerald* radioed the shipping company and said that she would be delayed in reaching the Soo Locks because of the bad weather conditions. Captain McSorley, probably thinking that the storm would stay mostly to the south, steered northeast toward the Canadian shore of Lake Superior. Sailors on Lake Superior had learned from decades of experience that the north shore offered some protection from the prevailing northwesterly winds. But now the wind was coming from the northeast, and McSorley turned southeast again.

The *Edmund Fitzgerald* had left Superior, Wisconsin, at 2:15 P.M. on November 9. A slightly larger carrier, the *Arthur M. Anderson*, left a nearby port at 4:15 P.M., also headed east, for the steel mills at Gary, Indiana. The two captains knew each other, and they agreed to stay in radio contact. The *Fitzgerald*, which was faster, pulled ahead of the other ship but later slowed down so as not to be too far out of touch. When the *Fitzgerald* was 2.5 to 3 miles southwest of Michipicoten Island, in Canadian waters off the eastern shore of Lake Superior, Captain McSorley told Captain Cooper of the *Anderson* that he was "rolling some" but would continue nevertheless. It was now November 10. At about 3:30 P.M., McSorley said that he had a fence rail down and had lost a couple of vents and was listing. He said he would slow down to allow the *Anderson* to catch up with him. Captain Cooper asked McSorley if his pumps were working, and the other master replied, "Yes, both of them." The *Anderson*'s radar showed that the *Fitzgerald* was 17 miles ahead of her and a little to the right.

The ships were nearing Caribou Island and the dangerous shoals near it. At about 4:10 P.M., Captain McSorley told Captain Cooper that he had lost both his radars and asked him to keep track of the *Fitzgerald* and plot its course.

About 4:39 P.M., McSorley spoke via radiotelephone with the Coast Guard station at Grand Marais, Michigan, and was informed that the radio beacon at Whitefish Point was not working. Whitefish Point is less than 50 miles from the Soo Locks, where McSorley might have expected to find shelter, but at that moment he was miles to the northwest. The *Fitzgerald* was next heard from between 5 and 5:30 P.M., when McSorley spoke with a Great Lakes pilot on board the westbound oceangoing Swedish freighter *Avafors*. The pilot said that the lighthouse at Whitefish Point was working, but the radio beacon, which McSorley needed for navigation, was still off. While the radio phone was open, the pilot heard McSorley saying to someone in the wheelhouse, "Don't let nobody on deck." The pilot thought that McSorley said something about a vent, but he could not be sure. McSorley then told the *Avafors* that his ship had a bad list, had lost both radars, and was taking heavy seas over the deck in one of the worst seas he had ever been in.

At 7 P.M., the first mate of the *Anderson* spoke with McSorley, giving him his position: 10 miles ahead and 1 to 1.5 miles east of the *Anderson*. About ten minutes later, he warned the *Fitzgerald* about northbound ship traffic about 9 miles ahead and asked how the ship was doing. McSorley replied, "We are holding our own."

Near 7:20 P.M., the *Edmund Fitzgerald* disappeared from the *Anderson*'s radar. About this time, the snow that had been falling and obscuring vision stopped, and the men on watch aboard the *Anderson* could see lights on shore about 20 miles away, but they could not see the *Fitzgerald*. Despite attempts to call the *Fitzgerald* on the radio phone, they made no contact. At 8:32 P.M., the worried master of the *Anderson* called the Coast Guard station at the Soo Locks. At the Coast Guard hearing a month later, Cooper testified that the Coast Guard radioman did not pay much attention to his concern and told him to look out for a 16-foot boat, a small vessel that had gone missing earlier. Cooper's recollection may have been faulty, however, for the log shows that the Coast Guard immediately tried to contact the *Fitzgerald* both on its own radio and through a commercial radio station in Michigan.

At 8:40 P.M., the Coast Guard concluded that the *Fitzgerald* was missing and notified the Coast Guard Great Lakes Rescue and Coordination Center in Cleveland, Ohio. The message worked its way through the channels of the bureaucracy, and at 9:15 P.M. the Cleveland center ordered an aircraft to leave the station at Traverse City, Michigan, and search the area where the ore carrier was thought to have disappeared. It is uncertain how effective an aerial search would have been in a violent storm on a dark night, but the Coast Guard had to try to locate the missing ship.

The nearest Coast Guard cutter, the 110-foot *Naugatuck*, was the logical choice to send out for the search-and-rescue mission. But the *Naugatuck* was down for maintenance. Hastily called back into action, she set out on the raging lake, but one of her oil lines failed, and she had to limp back to the base for repairs. It was 9 A.M. on November 11 before she was in shape to leave, and she did not arrive at the search area until 12:45 P.M. Any seamen who had managed to escape the sinking *Fitzgerald* would, of course, have been dead of hypothermia hours earlier, but the Coast Guard's mission included locating the wreck and recovering bodies. The *Naugatuck* was not even supposed to go out in such high winds and heavy seas, and her initial orders had been to proceed to the entrance of Whitefish Bay and wait until the weather had moderated.

At the Coast Guard's Duluth, Minnesota, station, there was a sturdier vessel, the 180-foot buoy tender and icebreaker *Woodrush*, which got under way by 12:08 A.M. on November 11, in half the usual preparation time. But the *Woodrush* had to cover 320 miles and did not reach the search area until twenty-four hours later. Several other Coast Guard vessels of varying sizes were pressed into action and took part in the search. The *Arthur M. Anderson* broke off her own voyage and turned around to look for the missing *Fitzgerald.*

Seven other lake ships were waiting out the storm in the sheltered waters of Whitefish Bay, but only two consented to go out onto the lake and join the search. The *Avafors* and two other oceangoing Swedish ships refused to turn back because they said conditions were too dangerous.

Despite the dangerous winds, the Coast Guard sent out two helicopters and a fixed-wing aircraft; one helicopter was equipped with a 3,800,000-candlepower searchlight. A third Coast Guard helicopter was sent from as far away as Elizabeth City, North Carolina. Canada also contributed planes to the effort. The search went on around the clock but without finding a trace of the *Edmund Fitzgerald.* Late at night on November 13, the intensive search was called off, though periodic survey flights continued until the end of the year.

On November 14, a navy plane carrying a magnetometer detected a large magnetic anomaly (translation: a big object made of iron or steel) on the lake bottom, slightly north of the Canadian boundary near Whitefish Point. The Coast Guard sent the *Woodrush* out with a side-scan sonar; the mission confirmed that there was indeed a big wreck there, though it could not be identified. Another search was made toward the end of November using a professional search-and-salvage company from Virginia, with the *Woodrush* as mother ship. Despite high winds and bad waves, the search team was able to identify the wreck as the *Edmund Fitzgerald.*

The same search company went out with the *Woodrush* again from May 12 to May 16, 1976. It used sonar to home in on the wreck and then placed an anchor line for a survey. The survey was conducted with the U.S. Navy's renowned ROV, *CURV III,* which could go as deep as 7,000 feet—far deeper than any spot in Lake Superior. *CURV III* measured approximately 6 × 6 × 15 feet, about the size of a small pickup truck with a cap over the bed. Armed with a still camera and two black-and-white video cameras, it performed twelve dives between May 20 and May 28, logging 56 hours of bottom time and taking 895 color photos and over 43,000 feet of videotape. On the very first day, one of the video cameras picked up the white letters of the name *Edmund Fitzgerald* on the wreck's stern, which was resting upside down on the lake bed 530 feet below the surface. The bow section, badly battered but right side up, was located 170 feet from the stern, pointing in a different direction. The middle section, approximately 200 feet of it, was missing. It lay scattered in bits and pieces.

Seaward, Inc., the Virginia-based salvage firm that had conducted the sonar searches, combined its results with *CURV III*'s and created a series of sketches of the wreck that was appended to the National Transportation Safety Board's final report.

No bodies were ever found, despite the intensive searches. But a few bits of flotsam were found shortly after the disaster by the commercial ships that had joined the quest. There were old-fashioned cork life jackets and pieces of life jackets, a stepladder, lifeboat oars, two propane tanks from the deck, and other small floating debris, plus the badly mangled lifeboats. One of the lifeboats was actually only a 16-foot section that looked as if it had been bitten off. Two empty life rafts were also found.

Three theories were current about the sinking, all reasonable under the circumstances. One held that the *Fitzgerald* had struck the notorious Six-Fathom Shoal near Caribou Island and sprung a leak that later sank her. One held that she had been hogged by a monster wave and perhaps broke in two on the surface. The third theory was that she had been flooded through her cargo hatches. The *Fitzgerald* had twenty-four cargo hatches. Unlike those on the *Marine Electric*, they were transverse, running across the deck from side to side rather than lengthwise. The hatch covers were one-piece and very heavy. For convenience in pushing piles of cargo around, the hold was one continuous space. It was not divided by watertight bulkheads, a serious deficiency.

A Coast Guard Marine Board of Investigation convened and heard testimony. Much of the testimony was far from confidence-inspiring. Captain Dudley Paquette, who had known Captain McSorley for years and whose own freighter had lain beside the *Fitzgerald* at the loading dock on November 9, told the board that McSorley had sailed with only a few dogs on the hatch covers engaged. It was a Sunday, and McSorley would have had to pay the crew time and a half for any work they did that day. The rest of the dogs could be tightened down the next day, for ordinary pay. This was common practice among lake captains in good weather, Paquette said, and he had done it himself. Management liked keeping expenses down, just as it liked keeping schedules and loadings up. But the storm that overtook the *Fitzgerald* did not give the crew time to dog the hatches. Paquette did not think much of McSorley but acknowledged that he was a competent mariner who would not knowingly put his ship or his crew in danger.

George Burgner, a retired cook from the *Fitzgerald*, testified that the ship had been sloppily built. The keel frequently came loose from the hull, and once, when the ship was laid up for winter repairs, a welder pushed a crowbar between the keelson and the hull plating (where there should have been no empty space) and shoved out several old welding rods that had been left behind. Further, said Burgner, Captain McSorley used to complain about the *Fitzgerald* having a "wiggly thing": that is, she flexed and twisted excessively in heavy seas. This kind of action is hard on steel and embrittles it, especially in cold weather. Burgner also claimed that the repairs on the keelsons were carelessly done. In some places, he said, the welders had simply tack-welded the keelsons back onto the hull instead of using a continuous weld. And Burgner said that he had once overheard two crew members tell McSorley that the keel was loose again. McSorley allegedly replied, "All this son of a bitch has to do is hold together one more year. After that, I don't care what happens to it." Burgner also said that McSorley was "afraid of his ship." During his career as a ship's cook, Burgner had plenty of opportunity to pick up gossip from crew coming to the galley to be fed, and for several winters he had an additional job on the *Fitzgerald* as a ship-watcher. A ship-watcher stays on the ship and keeps an eye out for thieves and vandals while the vessel is laid up for the winter, and he has plenty of time to chat with repair crews and gather choice items of information from them.

Although Burgner obviously thought poorly of McSorley, he had enough confidence in the captain to continue working under him until September or October 1975, when a foot injury forced Burgner to leave the ship. Some other seamen, however, thought Burgner was a know-it-all and a blabbermouth and did not put much faith in what he had to say. Neither did investigators from the historical societies.

But it was not only Paquette and Burgner who indicated that negligence was to blame for the sinking. The company's logs showed that McSorley held lifeboat drills only every couple of weeks instead of once a week as required. The *Fitzgerald* was not equipped with a depth finder, which would have helped her avoid shoals, or an inclinometer, or a trim indicator.[23] She did not have an EPIRB. At the time, Coast Guard regulations did not require these radio beacons for freshwater ships, and an EPIRB would not have saved the lives of the *Fitzgerald*'s crew—but it would have led searchers directly to the wreck.

The pictures taken by *CURV III* showed a gaping gash in the right bow of the *Fitzgerald* and some damage to the hatch coamings. The dents and gashes in the hatch coamings were there before the sinking—they were an unavoidable side effect of the loading equipment banging against them—and were scheduled to be repaired at the end of the voyage. The ROV showed no evidence of grounding on the upturned stern; the bottom of the bow could not be examined because it lay buried 25 feet or more in the mud. Unfortunately, *CURV III* stirred up such clouds of sediment that the pictures were not very clear.

The final report cited flooding through the cargo hatches as the most probable cause of sinking, given the violence of the waves. The missing fence rail and vents had probably been carried away by some heavy object on the deck that had come loose, perhaps the crane that lifted the hatch covers or one of the hatch covers themselves. The images taken by *CURV III* showed that many of the hatch dogs were undamaged and open, leaving the hatch covers vulnerable to being knocked loose. Their condition indicated that they had never been dogged down.

The *Fitzgerald* showed no evidence of having broken apart before sinking, nor did she capsize. Had she capsized, the loose and heavy

[23] In nautical parlance, "trim" refers to the relative positions of the bow and stern: in other words, whether the vessel is nose-heavy or tail-heavy.

cargo of taconite pellets would have torn the hatch covers from their fastenings, and there was no visible evidence of this. Instead, the board concluded, water had come into the cargo hold, sunk down through the taconite pellets, and accumulated. The bilge pumps would have been ineffective, for they would have quickly become clogged. The ship would have gone down bow-first, too fast for a *Mayday* call, and broken up either as she sank or when the bow hit the lake bottom. The hole in the bow would probably have been created by the impact, too.

There were dissents from the board's conclusions. The Lake Carriers' Association, a group of shipowners, maintained that the *Fitzgerald* must have struck ground, and Captain Cooper of the *Anderson* said repeatedly that the *Fitzgerald* had to have been over the shoals. (However, both the *Anderson* and the *Fitzgerald* were apparently zigzagging to cope with the waves, and the positions reported by the *Anderson* were inaccurate. The Coast Guard pointed out that, in order for the *Fitzgerald* to have been where Captain Cooper said she was at certain times, the *Anderson*'s speed must have varied from 5 miles per hour to an impossible 66 miles per hour.)

Whatever the cause, the loss of the *Edmund Fitzgerald* inspired a Canadian folk singer named Gordon Lightfoot to compose a song, which became a hit and is reproduced in the popular books on the *Fitzgerald*'s sinking. But it was not the last word.

After the Coast Guard's initial investigation, several expeditions explored the wreck. In late September 1980, Jacques-Yves Cousteau's research vessel *Calypso* was on the Great Lakes filming a documentary. One of Cousteau's top-ranking divers, Albert Falco, accompanied by another diver, went down in a submersible, one of Cousteau's renowned "diving saucers." Limited by their batteries, they had enough power for only about thirty minutes, which did not give them much time for observation.

The Frenchmen related how the sonar of their diving saucer picked up the wreck for which they were searching almost immediately. The bottom over which they were flying looked as if it had been worked over by a gigantic plow. Huge furrows 30 to 60 feet long radiated from the hull in all directions. Suddenly, they confronted the dim silhouette of the wreck. They saw a monstrous hull and a porthole. Startled, they thought they saw a light burning inside the *Fitzgerald*—it was a haunted ship! But it turned out to be the reflection of their mini-sub's floodlights off the glass of a porthole. They flew around the hull, "a bit like a helicopter circling a fortress," and found the hull in three pieces. The bow looked like a giant accordion, they thought. In places, it looked as if a giant had tried to dismantle it with a huge hammer. They were able to make out the *Fitzgerald*'s name twice as they flew slowly over the wreck. They passed the identification mast and the radar mast. They peered into the bridge, where they identified the helm, the gyrocompass, and a stainless-steel stair railing, and they thought they saw bodies. Falco commented that the cold, fresh water of the Great Lakes keeps bodies well. Then it was time to terminate the dive.

Based on their information, Jean-Michel Cousteau, the son of the famed undersea pioneer, concluded that the *Fitzgerald* had broken in two at the surface, but both sections had floated for a while before sinking. (How this jibes with Falco's observation that the hull was broken into three parts is not explained.) The gash in the bow, he thought, had occurred when the bow section was pushed by waves and wind against the stern section.

In 1989, Michigan Sea Grant sponsored an expedition using a very sophisticated ROV and obtained images of very good clarity. They showed, among other things, that the port door on the *Fitzgerald*'s bridge was open, suggesting that someone had either tried to leave the vessel or had at least thought about it. The damage to the bow would have been caused by the drag of the water on the way down and by the impact as it struck bottom. The hatch crane was finally located. The spilled taconite cargo was shown to be primarily on the starboard side of the bow, confirming McSorley's report of a bad list. Some was covering the deck; it had apparently spilled from the severed stern, which sank more slowly.

In 1994, there were two expeditions. One was led by the famous Canadian diver-scientist Dr. Joseph MacInnis as part of a six-week ecological survey of the Great Lakes and St. Lawrence River. MacInnis is renowned for having been the first person to dive under the Arctic ice at the North Pole, and cold-water exploration is his specialty. A man of many talents, he was originally trained as a physician but is also an environmentalist, oceanographer, explorer, poet, writer, and philosopher. One of his inventions was the *Sub-Igloo*, a transparent acrylic underwater habitat that he placed on the floor of Resolute Bay, 600 miles north of the Arctic Circle. From inside this habitat, MacInnis and his colleagues could observe firsthand the rich and fascinating marine life of the Arctic Ocean. But on the 1994 expedition, MacInnis was not looking for marine life. Inspired by Gordon Lightfoot's ballad, he wanted to make a television documentary of the *Fitzgerald* tragedy.

MacInnis used the R/V *Edwin Link* as his base, and he had a 22-foot, three-seater submersible, the *Celia*, which made six dives and shot ten hours of videotape. Accompanying Dr. MacInnis was Thomas Farnquist, executive director of the Great Lakes Historical Society. Farnquist, an amateur diver since the age of sixteen, began his professional career as a middle-school science teacher in northern Michigan. As a hobby, and as a device for gaining his pupils' interest, he plotted shipwrecks in Lake Superior near Whitefish Point. An expert diver, he became involved in the *Fitzgerald* affair two days after the sinking. He has made four manned submersible dives on the wreck, and in the summer of 2002 he went back with side-scan sonar on another search for clues. (He says that his new sonar is so sensitive that it can detect objects as small as a lump of coal only 3 inches in diameter.) He found that the spar deck had collapsed into the hull but gained no new insights into the cause of the sinking.

Farnquist concluded that the cause of the disaster could still not be determined, but his personal belief is that the ship, in a trough between waves, had struck a rock and hogged. In any case, the bow had struck the lake floor with tremendous force, causing catastrophic damage to the ship, scattering taconite pellets for hundreds of feet, and plowing up great chunks of the clay that coated the lake bottom, chunks so sharp-edged that they resembled rock slabs. Farnquist believed that the *Fitzgerald* had plunged downward in one piece at full momentum, with the propeller still turning. Twenty-nine thousand tons of ore, plus the weight of the ship itself, provided tremendous force. The momentum was such that the *Fitzgerald*'s bow penetrated clear through the clay of the lake bottom to the bedrock below. Farnquist also theorized that the rear third of the ship might have been above the water when the bow struck bottom, which could account for its being wrenched loose with such violence.

The other 1994 expedition was organized and financed by a Michigan businessman named Frederick J. Shannon, who used a powerful Canadian tug as his mother ship and a 16-foot submersible, the *Delta*, for his investigation. Shannon got images of excellent quality but discovered nothing new—except a human body in a life jacket. The body was assumed to be the remains of one of the *Fitzgerald*'s crew, although it could have been another drowned person who had drifted to the wreck site. The body was left where it was found. On his final dive, Shannon left a memorial plaque. The Canadian government, in whose waters the wreck is considered to lie (it is right on the boundary), declared it a grave site. Shannon believed, counter to most, that the *Fitzgerald* did break apart on the surface due to structural failure from the numerous stresses she had accumulated over her seventeen-year working life carrying massive loads.

In 1995, a special mission authorized by the Canadian government and directed by Joseph MacInnis sent a diver in an armored Newt Suit to retrieve the ship's bell, which is now part of a memorial at Whitefish Point. And on September 1 of that year, two American sport divers, using a special breathing mixture combining oxygen, nitrogen, and helium in six different proportions according to the depth, reached the wreck of the *Fitzgerald.* The dive took three hours in all: eight minutes for descent, four minutes on the bottom, and the remainder spent ascending in careful stages for decompression.

Metal lasts a long time in cold, fresh water, so it is likely that the wreck of the *Edmund Fitzgerald* will endure for many years. But the mystery of her sinking is unlikely ever to be solved.

* * *

Two ships, two tragedies, whose causes remain hotly debated. One may well be a massive cover-up; the other shows that even honest forensic effort cannot always arrive at the truth.

6 Human Error or Just Bad Luck? The *Arctic Rose* and the *Margaretha Maria*

At 7:33 in the morning on April 2, 2001, a computer in a Coast Guard mission control center in Virginia picked up an emergency signal from a sinking fishing boat an entire continent away. It was from the emergency position indicating radio beacon (EPIRB) of the trawler *Arctic Rose*, sinking in the frigid Bering Sea. The radio signal from the *Arctic Rose* was picked up by a satellite and instantly relayed to the East Coast center, which immediately flashed it to the Coast Guard command center in Juneau, Alaska, the closest such station to the scene of the disaster. Within two minutes of the original distress signal, the Coast Guard in Juneau was alerted. The Coast Guard immediately called the vessel's owner, Dave Olney, who was asleep in his bed in Seattle, Washington (4:33 A.M. local time). Alarmed, Olney frantically tried to contact the *Arctic Rose* by satellite phone—but in vain. The luckless trawler and its crew were already on their way to the sea floor, 428 feet below.

The signal from the *Arctic Rose*'s beacon indicated a site some 750 miles from the Coast Guard base at Juneau. A Coast Guard plane set off to search for survivors, a routine duty, even though it was almost a futile effort considering the icy chill and rough waters of the Bering Sea, where 20-foot swells are commonplace and not considered really bad by those who fish the area. It reached the site given by the emergency beacon at 8:30 A.M. The Coast Guard pilot found only the body of the skipper, Dave Rundall, who hailed from

Hilo, Hawaii. Rundall was wearing a survival suit, colored bright red for visibility. It had not helped him. The suit was unzipped and filled with seawater—apparently, he had pulled it on after he had jumped off the sinking ship—and Rundall was cold and lifeless.

His body was pulled from the sea by the *Alaskan Rose,* a sister ship fishing about 10 miles away and guided to the site by radio from the Coast Guard search plane. Later another body without a survival suit was detected floating in the water but could not be recovered because of the high waves. Five more survival suits (all empty), an oil slick, and an untenanted life raft were also found. In all, fifteen people were lost; the sinking of the *Arctic Rose* was the worst disaster in the area since 1982, when the Japanese trawler *Akebono Maru* capsized and took thirty-two crew members to their deaths. It was also the worst disaster for an American fishing vessel in fifty years.

Some experts had grave doubts that the *Arctic Rose* should have been fishing in the Bering Sea at all. Amateurishly built in a small Mississippi boatyard by a Vietnamese refugee entrepreneur, she was originally designed as a shrimp boat. (After the tragedy, the Coast Guard learned that the builder had not even used plans.) She was sold and converted into an Atlantic scalloper for use off the New England coast. Still later she was converted again, this time into a trawler to work off the coasts of Washington and Oregon. Shabby and ill used, she was spotted laid up in Seattle in late 1998 by her last owner, Dave Olney, who bought her in 1999 and spent considerable money restoring her to working condition and converting her into a factory trawler (a trawler that processes the fish on board as they are caught). But the conversion created problems with the boat's stability, even though Olney added 7 tons of lead to the keel.

Olney engaged a Seattle consulting firm to work out a set of figures for maintaining the *Arctic Rose* on an even keel; the calculations were complex, involving the amount and distribution of fuel in the fuel tanks and the amount of fish caught, all of which had to be constantly recalculated as the ship used up fuel and hauled in fish. Too heavy in the bow and the ship might take a nosedive like the *Derbyshire.* Too heavy in the stern and a following wave might flood the interior through the fish hatch. Unfortunately, Bering Sea fishers tended to forget these safety rules under the pressure of hauling in a catch and later processing it.

At 92 feet of hull length, the *Arctic Rose* was considered unsuitably small for the Bering Sea: not really big enough for the waves and winds, and too small to hold a really profitable cargo of processed fish. On some trips, the earnings from the catch were only half of what the crew and the owner had hoped for. As a result, crew members, whose pay is based on the earnings for each voyage, seldom stayed for long. There was a steady turnover of captains, mates, engineers, deckhands, and the workers who beheaded and gutted the fish. Nevertheless, Dave Olney kept the ship operating each fishing season.

In January 1999, Olney hired a new captain, David Rundall. Rundall, aged thirty-four, was acknowledged to be a competent captain, but according to newspaper reports he was overbearing, hotheaded, and verbally abusive, with a macho disregard for dangerous situations. (Rundall's wife loyally characterized him as very safety conscious. She also said that "he had a bit of a temper ... but he wasn't abusive. He was, I guess, what you'd call brutally honest.")

On the *Arctic Rose*'s last voyage before she sank, the engineer had a heated confrontation with Rundall to persuade him to run for shelter behind an island instead of bulling his way through a storm to reach the fishing grounds on schedule. The engineer quit on the ship's return to port, bitterly disappointed with his meager earnings. His place was taken by Olney's younger brother, Mike, who moved up from assistant engineer.

Life on an Alaskan fishing boat is lived under high pressure. When the fish are running, netting them, hauling them in, and processing them before they spoil goes on around the clock. The processing crews work six hours on, six hours off. Time is money, and everyone is aware of it. Any mechanical failure can cost thousands of dollars in lost time and catch. Morale aboard the *Arctic Rose* on this voyage had been low. The weather was bad, and there were few fish. The day before the sinking, Rundall had needed to send a young crewman out on deck with a baseball bat to break off the ice that had formed on the ship's rigging and superstructure. A heavy coating of ice can cause a ship to capsize and is a danger for most of the year on the Bering Sea.

On the afternoon of April 1, 2001, Rundall radioed his friend John Nelson, the mate of the *Arctic Rose*'s sister ship *Alaskan Rose*, which was fishing about 10 miles away. Rundall was frustrated and in a foul mood, Nelson later said. Fish waste had clogged a pump in the processing room, forcing operations to stop. But later Rundall radioed his friend again with better news: the pump was fixed, the production line was up and running, and he had found a huge school of valuable flathead sole. In fact, he hauled in 20,000 pounds. Rundall and Nelson had another radio chat at 10:30 that night, the last conversation the *Arctic Rose* had with anyone. Before dawn the next morning, she was gone. In a sad twist of fate, it was Nelson who pulled Rundall's body from the water.

Even before a Coast Guard board of inquiry could be convened, people were circulating theories about the cause of the tragedy. Some said the boat must have capsized, trapping the crew below deck. Others thought that the much-altered structure of the boat had not been equal to the stress upon it and had simply broken in half. These speculations were pure conjecture. The Coast Guard needed something more solid on which to base an investigation. As early as April 12, Lieutenant James Robertson of the Coast Guard arrived in Dutch Harbor, on the island of Unalaska, one of the Aleutian chain that curves out into the sea for some 1,100 miles from the tip of the Alaskan Peninsula. The seasonal center for the Bering Sea fishing fleet, Dutch Harbor was the logical place for Robertson to begin a marathon session of interviews. Before he was finished, he had spoken with suppliers and shipyard workers who were familiar with the *Arctic Rose*, with fishermen at the local bar, and with all twenty-seven crew of the *Alaskan Rose*.

While Robertson was thus engaged, other Coast Guard investigators interviewed executives at the Seattle company that owned the missing trawler and tried—in vain—to locate the man who had built the *Arctic Rose* thirteen years earlier in his ill-equipped Mississippi boatyard. At Coast Guard headquarters in Washington, DC, experts constructed a computer model of the *Arctic Rose* to run through various scenarios of the sinking. According to *Online Mariner*, an

Internet trade paper, the Coast Guard even checked out the possibility of collision with surfacing submarines (both U.S. and Russian), stray icebergs, freak waves, and even falling space junk. These last investigations yielded no results.

However, former crew members told of the *Arctic Rose*'s instability and mechanical problems. The bearings of the propeller shaft sometimes overheated, posing the risk that the shaft might seize and damage the bearings, causing them to leak. Olney shouldered the expense of replacing one bearing and had custom-made temperature gauges installed on all of them, but the problem did not go away. Then, too, the pumps in the processing room sometimes got clogged, and the floor of that large room would flood, sometimes over the crew members' boot tops. Even 6 inches of water on the floor of the processing room could weigh several tons. And this immense weight, sloshing back and forth as the boat rolled and pitched with the waves, could easily destabilize it. Furthermore, as the boat's center of gravity shifted while the nets were being hauled in over the stern, the engineers were supposed to pump fuel from one tank to another to correct the trim. Was it possible that they had miscalculated and allowed their boat to turn turtle?

The Coast Guard needed concrete evidence to confirm or disprove these theories. The Marine Board of Investigation chartered a ship of convenience, the 155-foot trawler *Ocean Explorer*, which the National Oceanic and Atmospheric Administration (NOAA) had been using in a survey of how much damage bottom-trawling inflicted on the sea-floor fish habitat (thus affecting future catches). Already on board was a camera-equipped ROV owned by Maritime Consultants of Puyallup, Washington, which was placed at the Coast Guard's disposal, and a towed-sonar array. The *Ocean Explorer* was not the ideal ship for investigating a wreck: it had no thrusters and therefore had no way of maintaining a stationary position, which made things very difficult for the ROV pilot—but it was there on the spot.

The Coast Guard searchers already knew the approximate location of the wreck, calculating it from the EPIRB signal, the oil slick, and the floating debris. Using the towed sonar, they found the sunken trawler in three passes, an unusually short time. The *Arctic Rose* was lying about 2.5 miles from the position of the EPIRB; she had traveled a good distance while sinking. Did this mean she had remained partly buoyant? The answer to this question could have a bearing on the cause of the sinking.

The ROV, about the size of a laundry dryer, was launched on July 18 to document the wreck. The video record, captured by the computer on the mother ship, showed an endless expanse of flat, sandy-silty bottom with small rises and hollows, inhabited by a scattering of ghostly white crabs. Now and then a fish would swim past the camera, its sides gleaming in the ROV's floodlight. Then a large, dark object hove into view: the hull of a ship! The camera picked out the outlines of a bow and a wheelhouse and a series of white letters spelling out *Arctic Rose* against the chipped blue paint of the hull. Then the ROV got tangled in several loose lines trailing from the hull. A cable had gotten wound around the vertical thruster. The camera showed what looked like a tangled, writhing mess of colorful spaghetti as the operators tried in vain to work the ROV free. Then the umbilical cable broke, and the $100,000 machine was lost. Two important points were established, however: the vessel on the sea floor was the *Arctic Rose*, and it was intact, which meant that the ship had not broken up either at the surface or during its descent. It was resting on its keel and the starboard bilge keel on the soft sediments.[24]

The Coast Guard, which had already dedicated $200,000 to the expedition, asked for more money. It was forthcoming, and a second, larger ROV, supplied again by Maritime Consultants, was sent down on August 24. This time the ROV operators kept their little remote-controlled vehicle pretty well clear of entanglements—not entirely, however, for in attempting to cut the first ROV free, the second became entangled itself. Since the *Ocean Explorer* lacked bow thrusters and could not hold a fixed position, it was forced to steer in circles above the site of the *Arctic Rose*, greatly complicating the operators' task. Finally, the captain gave orders to drop anchor, which held the mother ship steady enough for the ROV pilots to work their 76-inch × 36-inch craft free.

[24] A bilge keel is a row of finlike plates that runs along the length of the hull at the point where the vertical sides of the vessel meet the rounded shape of its bottom. Its function is to inhibit rolling.

The Coast Guard's video showed a good deal of marine life in addition to the wreck itself. Clouds of tiny white creatures swam and jittered past the camera; an occasional octopus jetted by, tentacles trailing behind it. A few slender, spindle-shaped organisms—squids perhaps—darted by. A large gray fish with fat white lips peered sulkily out from beneath the hull. But it was the wreck, not the biota, in which the Coast Guard was interested. The camera slowly panned along the hull, peering inside the open rear hatch through which the nets were paid out and hauled back in. This door was supposed to be kept closed to keep the sea out, but the crew, said witnesses, liked to tie it open even in cold weather to escape the stench from the processing tables.

The interior—what could be seen of it—was deserted. Again the video recorded the *Arctic Rose*'s name on the bow as proof that this was the wreck in question. The propeller, half buried in the sediments, appeared to have a line wound around it, though what connection this might have to the sinking was not clear. The rudder was hard over to the left, as if the helmsman had tried to make a sharp turn—to face a wave head-on? Images from the towed sonar showed netting spread over the bottom like a ragged shadow on the starboard side, an indication that the *Arctic Rose* had turned over on that side before sinking.

Maintenance on the *Arctic Rose* had evidently been minimal. There were many patches where the blue paint on the hull above the waterline was flaking off, exposing the white undercoat. The ROV operators found a large, irregular patch below the waterline of the sunken trawler that appeared to be rotten with rust, but, when they tried to punch through it with the ROV's mechanical claw, they found solid steel underneath. Apparently, the *Arctic Rose* had been structurally sound when she went down. The windows, unbroken, were another indication of structural soundness.

During the summer of 2001, a parade of witnesses testified about the *Arctic Rose*. She was top-heavy and stern-heavy because of Olney's modifications, said the consultant who helped Olney work out routines for safe handling, but a Coast Guard stability expert said that the trawler met the minimum requirements. More puzzling evidence was forthcoming about the weather. The mate of the *Alaskan*

Rose said that at the time of the sinking conditions were relatively calm, with waves of only 8 to 12 feet. At some point during the night, he said, the wind picked up and blew at over 40 miles per hour but soon subsided. He did not consider it a "serious storm." But a weather expert from the National Transportation Safety Board (NTSB) said that a weather front might have crossed the area where the *Arctic Rose* disappeared at just about the time of the sinking. This local disturbance could have created waves big enough to endanger the trawler.

A former crew member, Nathan Miller, told the board of inquiry that safety training on board the *Arctic Rose* consisted of trying on a survival suit.[25] The chief investigator asked him about the muster stations, to which the crew were supposed to report for evacuating the ship. Miller had to ask him to define the term.

"Where were you supposed to go in an emergency?" asked the investigator.

"I don't know. We were never told" was Miller's honest reply.

His desire for adventure on the high seas being kindled by programs on the Discovery Channel, Miller said he had pestered Olney for a job that January. He was hired as an entry-level fish processor. But he soon became disillusioned and apprehensive about the way the ship was run. The *Arctic Rose*, he said, habitually took on water in the processing room; it was often over his boot tops. Pumps frequently clogged, and no one seemed to be in charge of keeping the drains unclogged. In rough weather, more water could come in through the drain than went out. Watertight doors were sometimes left open, and the door between the processing deck and the engine room, which was made of cheap, hollow-core wood, was not watertight (though it was supposed to be). It was frequently tied open to ventilate the engine room, which blocked the escape passage from the processing deck to the main deck. Alarms on the ship went off every night, but the foreman told Miller not to worry about it.

---

[25] Miller's testimony lasted a full day and ran to 200 pages. The court-reporting service charges $900 for this day's material, which is far beyond the means of any ordinary author. This author was told that a major television company, interested in making a documentary on the *Arctic Rose*, found the document too expensive.

Miller also told of Dave Rundall's bad temper and abrasive personality and went on to say, "I also saw him lose his cool and make some bad decisions that affected all of us."

The former chief engineer, Milosh Katulich, recounted how Rundall, intent on making time to the fishing grounds, had insisted on motoring ahead at top speed through bad weather, with waves that caused the whole boat to vibrate when they struck it, stressing the engine and the vulnerable propeller shaft. He had to confront the headstrong skipper to get him to change course and head for shelter.

A former captain, Jim Kelley, said that he rarely took the *Arctic Rose* far from land because the boat handled poorly. Another former captain, Mike Hastings, gave his opinion that the 92-foot-long *Arctic Rose* was too small for the severe winds and waves of the Bering Sea. "I slept with one eye open pretty much," he said. Olney himself, who often skippered the boat in person, claimed that she "rode like a little duck."

Another witness, a veteran fisherman, said he thought that a combination of factors was probably involved in the *Arctic Rose*'s demise. The ship was known to be stern-heavy. An unnoticed leak into the area where the steering machinery was located could have brought the stern so low that waves would have surged in, flooding the factory deck and then the freezer hold. There would have been only one escape route from the crew's cramped quarters, just forward of the factory area: up a narrow staircase to the wheelhouse and then out the wheelhouse door. The ship would have sunk so quickly that the crew, exhausted and asleep, would not have had a chance. The captain and the mate, who bunked on the upper deck behind the wheelhouse, had a better chance of saving themselves, but neither survived.

In the face of this adverse testimony, Olney told the press that he wanted to testify himself but on the advice of his lawyer would plead the Fifth Amendment instead. Since the Fifth Amendment protects a person from having to testify against himself or herself, outsiders tend to presume that the pleader has something to hide. Olney's lawyer, however, said that he simply wanted to protect his client against overzealous government prosecutors who might try to nail him on charges of criminal negligence. The word *unfounded* was implied. (Five of the families of dead crewmen brought civil suits against Olney's company; as of this writing, one had been awarded $200,000.)

Crewmen who had worked for David Olney uniformly spoke well of him. Olney, a native of Seattle, had made his first fishing voyage at the age of seventeen. Only a year later he bought his first fishing boat. In time, he acquired his own company, Arctic Sole Seafoods. In 2001, at the age of forty-eight, he owned two fishing boats: the little *Arctic Rose* and its 120-foot sister ship, the *Alaskan Rose*. Olney often captained his boats himself, working side by side with his crew, and became close to them.

At a third hearing, held in November 2001 in Seattle, Olney, having obtained immunity from prosecution, testified before the Coast Guard's Marine Board of Inquiry. As he spoke, he burst into tears and had to leave the courtroom briefly. Pressed about safety procedures, he admitted that his crew were given instructions about safety, loads, and fueling by word of mouth rather than from a manual. Although the crew had changed the way they handled fueling, the consultant firm that prepared the manual had not been asked to revise it. As for the watertight door at the stern, Olney said that the crew knew it was supposed to be closed at all times, but they frequently asked to have it open on hot days.

At this hearing, a meteorologist presented a hindcast of weather conditions at the time of the sinking. His extrapolations indicated that the unfortunate *Arctic Rose* had been caught at the junction of two storm systems, which would have generated winds and waves severe enough to overwhelm a ship with a sleeping crew. Moreover, the wind could have shifted and buffeted the boat from several directions. The meteorologist told the board that the *Arctic Rose* had probably been hammered by waves with a maximum significant height of 24 feet, coming eight to twelve seconds apart. He also estimated that there had been a 45-knot wind coming from the southeast. Apparently, this was a very intense and local storm, over quickly. But experience has shown that most fishing vessels are sunk by local storms that are sometimes very limited in area.

Experts from the Society of Naval Architects and Marine Engineers (SNAME) spent months performing stability calculations with the Coast Guard. They modeled the motion of water in flooded spaces at various degrees of flooding to see how these would have affected the stability of the ship. They considered the effects of wind and waves, calculating that it would have taken a 36-foot wave or a 100-knot wind to have capsized the little trawler. They created nineteen different scenarios, ranging from slow leaking through the propeller shaft or other through-hull fittings to snagging the trawl nets on the bottom to flooding through an open door or hatch. Of these, fifteen were dismissed as unlikely or very unlikely. Slow leaks were ruled out because the ship was fitted with bilge alarms, and there was no indication that these had been triggered. The most likely cause was progressive flooding through the open trawl hatch and the aft door to the processing space. The water would have poured into the engine room and the processing space, causing the ship to capsize. It would have taken only a 10-foot wave to flood the trawler, which sometimes traveled with as little as six inches of freeboard.

In nautical engineering terms, *capsizing* means that the vessel has heeled over so far that it cannot be righted without outside help. If the watertight doors had been closed, as they were intended to be, even capsizing would not have sunk the *Arctic Rose* immediately, for the processing space would have provided enough buoyancy to keep her afloat until waves forced open a hatch or broke a window. Witnesses at the Coast Guard hearings testified that Captain Rundall was continually after the crew to keep the doors closed but that they always opened them again when his back was turned.

Chronically underfunded, the Coast Guard was unable to commission the mapping of the *Arctic Rose*'s debris field. The result was that the Board of Inquiry had to rely on the sketchy videos taken by the ROVs and the testimony of people who had served previously on the ship, supplemented by the expert consultants' calculations.

It was established that the watertight door was open when the wreck was found, but no one could say for sure whether it had been open to begin with or had come open as the vessel sank. The experts' calculations showed, in any case, that the *Arctic Rose* had probably capsized between one minute, forty seconds and two minutes, forty-three seconds after the progressive flooding began and had taken only four to eight minutes to sink after flooding began. The crew of the *Arctic Rose* never had a chance.

* * *

The *Margaretha Maria* was a small British trawler working the western end of the English Channel, where she disappeared without a *Mayday* call sometime between November 11 and 17, 1997.

Built in the Netherlands in 1958, she was sold to British owners in 1976. In 1994, she was bought by new British owners and underwent a thorough safety inspection and extensive repairs. The thickness of the steel plates of her hull was checked by ultrasound; the insulation on the electrical wiring was checked; the main engine, auxiliary engine, pumps, and gearbox were all overhauled. To improve her stability, 1.6 short tons of ballast were added under the fish hold, about midway between stern and bow. In late 1996, the aged but still functional trawler was treated to more renovation, including rebalancing and truing up the pitch of the propeller, in addition to relining the worn-out brake bands of the winch that paid out and hauled in the nets. Bad spots in the deck were repaired by welding. In short, the *Margaretha Maria* was declared fit for another two years.

Just under 71 feet long, the *Margaretha Maria* was a twin-beam trawler. On this type of ship, a long arm called a *derrick* extends out from each side of the boat and drags the nets along the sea floor by a long wire cable at its outer end. The nets themselves are made of heavy plastic mesh and are kept open at their mouths by a heavy wooden beam and floats. A shoe at each end of the beam allows it to slide over the sea floor and keeps it from digging in. A chain mat between the beam and the net disturbs the fish and, in the words of the Marine Accident Investigation Branch (MAIB) report, "encourages them to rise and swim into the mouth of the net." The underside of the net is shielded by matting to minimize damage to the seabed; how well it really works is conjectural.

As the net is towed along, the fish, disoriented and frightened, swim toward its closed end, known to fishers as the "cod end." A line called the *lazy decky* runs from the inboard shoe of each beam to another line that encircles the entrance to the cod end of the net. After each run, the towing cable is winched in, and the derrick arms are swung up by mechanical power until the beam breaks the surface. Then a device called the *gilson* pulls on the lazy decky, which in turn pulls the cod end closed. The whole arrangement is then swung onto the deck, and the cod end is opened to pour the fish out. Empty, each set of the *Margaretha Maria*'s beams, nets, and mats weighed 2.2 short tons—quite a load to lift in itself, and the weight of the catch would make it even heavier.

The trawler's new owners had given her longer derricks; they measured just under 28 feet. One can imagine the leverage they exerted on the ship while being raised to the vertical position for hauling the nets on board. If the load on the nets was not evenly balanced, the boat could capsize.

A trawler like the *Margaretha Maria*, which the Marine Safety Agency limited to short trips, needed only a small crew: a captain, a mate, and two deckhands, one to tend each net and derrick. Unlike the *Arctic Rose*, the *Margaretha Maria* had an experienced crew. The captain, aged forty-three, had formerly served on her as mate. He was certainly no young, macho hothead. The mate, aged forty-five, had served on the ship for almost two years. One of the deckhands was his twenty-four-year-old son, who had served on three trips. The other deckhand, aged thirty-three, had been with the *Margaretha Maria* since June 1997.

The *Margaretha Maria* had made three successful trips in the same area just before her sinking. There was no reason to suspect trouble was in store. In anticipation of another successful catch, the ship loaded up with 3.3 short tons of ice to chill the fish and 3,434 gallons of diesel fuel. Her very high frequency (VHF) and medium frequency (MF) radios and her cell phone were in good working order.

The little trawler chugged off from the port of Newlyn, near the tip of Cornwall, to the south-southwest at 4:30 P.M. on November 11. At that time of year, it is evening in Britain. The captain did not tell the owners just where he would be fishing (fishers are often reluctant to reveal where they go for fear that others will follow them and hone in on the catch), but they expected that he would return to the same productive area. The ship was expected to return to port on November 17 or early in the morning on November 18, although the crew had said they expected to be back by November 14.

The captain and one of the crew called home on the cell phone that evening. (Someone tried to call back to the ship later but could not make contact.) The call to shore at 7:15 P.M. was the last word heard from the *Margaretha Maria.*

The days passed; when the anxious owners had not heard from their trawler by November 17, they called the Marine Rescue Coordination Centre (MRCC). On November 18, when the *Margaretha Maria* was definitely overdue, the MRCC sent repeated radio and cell phone messages to the missing trawler. It got no response. Concerned officials called on nearby France for aid. Broadcasts in French were soon on the air. Harbors in southwestern England, southern Ireland, and northwestern France were searched for the missing vessel. Spain joined in the effort, broadcasting *Mayday* alerts to the Spanish fishing fleet. The results were nil. Shortly before noon, British search planes, a helicopter, and a seagoing tug combed a wide area of the sea. They found a few floating objects but nothing that could be linked to the missing *Margaretha Maria.* The search was continued for a few more days before being called off.

The first trace of the missing vessel turned up on February 18, 1998, when another British trawler pulled up a body in its nets about 50 miles south-southwest by south of Lizard Point, near the tip of Cornwall. Since the body was found in French territorial waters, it was turned over to French authorities. When it was identified as the captain of the *Margaretha Maria,* the MAIB called in the British Navy. The body's location gave a fair clue as to where the wreck might be located, although bottom currents could have carried it miles away. (In fact, the body turned up less than 5 miles from the wreck.)

A British mine-locator ship was dispatched to the area on February 27 and, with the aid of sonar, found several possible targets on the bottom. One looked promising—but bad weather, so often the villain in sea searches—curtailed the expedition. Not until March 17 did the navy send out another search vessel, this one equipped with an ROV. The ROV (later referred to in Parliament as the "yellow submarine," stealing a phrase from the popular Beatles song) was able to get enough video footage to identify the wreck as the *Margaretha Maria* but then got tangled in lines from the wreck and had to be abandoned.

The expedition yielded no information on the cause of the sinking, so the families of the lost crew members financed a survey of the wreck in April—probably with contributions from the public, for fishers are generally not wealthy. The families generously shared their video material with the MAIB, which concluded that it needed more information. For a week in June, the MAIB mapped the seabed and the wreck with side-scan sonar, then sent down an ROV to take pictures. This time the ROV became entangled twice and was left on the sea floor. (The contractor who owned it came back a few weeks later and retrieved it.)

Despite the abandonment of the ROV, a great deal of information was gathered: the trawler was sitting on soft sediments, tilting slightly to the right, about 360 feet below the surface. She had been fishing when she sank, for the nets were deployed, and the upright position of the derrick arms indicated that the crew were hauling in the nets. There was some damage to the hull, as might be expected from the stresses it suffered while sinking, but the ROV's camera picked up no tears or fractures in its steel plates. The derrick arms and the goalpost gantry to which they were connected were twisted, and the radio mast on top of the goalpost gantry was broken off.

The big surprise was the contents of the nets. The starboard net was nearly empty, but the port net was stuffed with shells and sand. The video taken by the first expedition showed an amount that experts on shore calculated to weigh 4.13 short tons. It appeared obvious that the weight of the shells and sand, magnified by the leverage of the derrick arm, had turned the boat on its side when the crew attempted to winch the load in.

The video also showed three watertight doors open. Why they should have been open in the raw November weather of the English Channel is a mystery, but it is likely that by letting in the sea they accelerated the sinking of the little trawler. Another surprise was the lifesaving equipment's failure to work. One of the vessel's two life rafts was found draped around the front of the wheelhouse, its cable fouled on the port derrick arm. The other was still in its canister, sitting on the sea bottom beneath the starboard derrick arm.

The *Margaretha Maria*'s rafts were equipped with up-to-date hydrostatic release units (HRUs), triggered by the rapid increase in water pressure as the vessel sank. Released by the HRU, the canister was supposed to pop open and the raft to inflate. As the raft became buoyant, a special weak link connecting the raft's painter, or rope, to the ship was supposed to break, allowing the raft to pop up to the surface. The HRUs had functioned as intended, but the weak links had failed to break. Why? The videos showed that the painter of the port raft had fouled on the derrick and other gear so that the raft, even though it had inflated, was pulling on the derrick instead of on the weak link. The painter of the starboard raft had gotten fouled in a mess of netting by the derrick so that there was not enough force on the line to make the canister open. But even the failed life rafts gave a clue, for their position indicated that the *Margaretha Maria* had sunk rapidly, stern-first. This near-vertical plunge was what caused the rafts' painters to become fouled.

The EPIRB could not be found. Its storage box was open, so it had apparently been released, but it had never emitted a radio signal. The failure could have been caused by a weak battery, but the battery had been checked four or five months before the accident and found to be in good condition. The MAIB concluded that the EPIRB, floating out of its box, had probably gotten trapped as the vessel sank. Water pressure might then have wrecked its mechanism.

A surprising conclusion that the investigators drew from this sinking was that it happened mainly because the winch was too powerful. A less powerful winch would not have been able to lift such a heavy load of shells and sand, and the boat could have remained manageable. As a result, the MAIB recommended that a study be made of stability under actual working conditions using a number of beam trawlers and that the findings "be made known to the industry as a matter of urgency." As a final item, the MAIB recommended to the Marine and Coast Guard Agency that it apply the results of the study of the free-floating performance of life rafts to EPIRBS, to ensure that these emergency beacons in fact self-release the way they are supposed to.

* * *

Unfortunately, the best recommendations help only when people put them into action. Hopefully, the tragedies of the *Arctic Rose* and the *Margaretha Maria* will send a sharp warning to the fishing community and to the people who design the boats in which the fishers go out and in so doing save other lives.

7 Intrepid, Independent Investigators: The Man Who Owns the *Lusitania* and His Peers

The cases discussed so far in this book represent deep-sea investigations that have been commissioned by one authority or another. However, there are a few stubborn and well-financed individuals who have persevered on their own in investigating and locating shipwrecks of interest. One is a business executive; two are graduates from the world of advertising and public relations, one of them a best-selling author and the other a treasure hunter-cum-amateur archaeologist. What they have in common is strong curiosity, unusual drive and energy, and a sense of adventure.

Gregg Bemis is an articulate, dynamic man in his early seventies. Born in St. Louis, Missouri, and raised in the Boston, Massachusetts, area, Bemis has always loved the sea. The son of a noted yachtsman who founded a sailing trophy for young people, he spent his early years in and on the ocean. After serving in the U.S. Marine Corps during the Korean War, he went into business, rising to the top management of three Fortune 500 companies. During his business career, he founded over forty new enterprises. One of these, an ocean-related venture capital firm, led him eventually to the wreck of the *Lusitania*. Bemis is currently the chairman of The Ocean Corporation, one of the world's leading schools for commercial divers, and was also chairman of Deep Ocean Engineering, a maker of ROVs and submersibles. He is a member of the Society of Naval Architects and Marine Engineers (SNAME) and a director and founder of ShipREX International, Inc.

In 1968, Bemis and two partners were working on the development of a portable saturation-diving system that could be used on deep-water salvage jobs almost anywhere in the world. One of these partners, looking for a suitable site to try out the invention, chose the wreck of the famous *Lusitania,* a ship whose sinking by a German submarine in 1915 helped spur the entry of the United States into World War I.

The *Lusitania,* a speedy luxury liner, was the pride of the Cunard Line. Built in Glasgow, Scotland, and launched in 1906, she measured 785 feet in length, with a beam of 87.8 feet. Her coal-fired boilers fed high-pressure steam to four huge turbines, which together generated 68,000 horsepower, and her four propellers gave her a cruising speed of 25 knots. Following the *Titanic* disaster, the *Lusitania* was fitted out with additional lifeboats, for a total of forty-eight. Her passengers, even the third-class ones, felt pampered and secure.

But the *Lusitania* was no ordinary liner. In return for a generous loan of £2,600,000 from the British government, Cunard had "*Lucy*" and her sister ship, the *Mauretania,* designed to be convertible on short notice into armed merchant cruisers. Each had built-in mounts for twelve deck guns capable of hurling 6-inch shells, and their officers and at least half the crew were required to be naval reservists. But the *Lusitania* was sporting no guns when she sailed on her final, fatal voyage.

She departed from Pier 54 in New York Harbor on May 1, 1915. On board were 1,257 passengers and 702 crew members to wait on them and keep the ship running. The cargo manifest listed foodstuffs (the German submarine blockade had caused a shortage of staples in Britain), metal rods and ingots, and a consignment of rifle cartridges.

The voyage was uneventful until the great liner neared the coast of Ireland on May 7, entering British waters (Ireland then belonged to Britain). German submarines were known to lurk there, primed to sink British ships and those of Britain's allies. Captain Turner ordered all lookouts to be doubled and the lifeboats to be swung out in preparation for emergency launching. The bulkheads were ordered closed to prevent water from flooding more than one compartment in the event a torpedo blew a hole in the hull. Steam pressure was to be kept high to give the ship enough speed to evade an attacking U-boat. But Captain Turner ordered the speed reduced to 15 knots because of fog, then raised to 18 knots when the fog lifted. The captain wanted the *Lusitania* to reach her destination of Liverpool, on England's west coast, at high tide the next day—May 8—rather than wait offshore until the tide was right. And he did not follow the Admiralty's standing orders to steer a zigzag course in order to make the ship a more difficult target. Just how much speed the captain could have gotten out of his huge ship is conjectural in any case, since six of her sixteen boilers had been shut down to conserve fuel.

Lurking off the Irish coast was the German submarine U-20, commanded by Captain-Lieutenant Walther Schwieger. The thirty-year-old Schwieger was something of a submarine ace, with many kills to his credit, including three on his current voyage. With only two torpedoes left, he was about to return to his base at Kiel to rest his crew and replenish his supplies. Peering through his periscope, he saw a large, four-stacked steamer in his sights—obviously a liner. Under the rules of war, civilian ships were not to be attacked without warning so that the passengers and crew could get off safely before the ship was destroyed. But Germany, under the spell of its bloodthirsty and egomaniacal Kaiser (Emperor) Wilhelm II, had announced in April that it would attack, without warning, any vessels flying the flag of the British Empire or any of its allies. Schwieger chose to follow the kaiser's directive.

As the *Lusitania* approached the fingerlike peninsula called the Old Head of Kinsale, Schwieger let loose one torpedo. It hit the liner just behind the bridge with a violent explosion. As terrified passengers struggled to orient themselves, a second, even larger, explosion followed. Schwieger noted in his log that the second explosion was probably caused by coal dust in an empty bunker that was ignited by the explosion of the torpedo; by ammunition; or perhaps by a shipment of high explosives in the cargo. (If there had been explosives in the cargo, Schwieger's attack would have been justified according to the laws of war.)

The explosions had blown such huge openings in the hull that the doomed *Lusitania* began to sink almost immediately. She heeled over sharply to starboard, so that the lifeboats on the port side could not be launched. Only a few of the starboard lifeboats could be launched because they had been secured to the deck with chains for safety reasons. The captain, trying to slow the liner's forward and downward momentum, ordered the engines into full reverse. But the assistant third engineer, panicking, brought the reverse turbines online before the propellers had stopped turning. Their momentum drove the turbines the wrong way; steam pressure built up, valves blew, and steam lines began rupturing. The captain ordered the helmsman to turn toward shore, but one of the burst steam lines was the one that powered the steering mechanism, and the rudder did not respond. The *Lusitania*, helpless, sank within eighteen minutes.

Of the 1,959 passengers and crew on board, 1,195 perished, including 123 Americans. The Germans and their Austrian allies exulted: at last they had given their haughty British cousins their comeuppance! But in the United States, the reaction to the sinking was outrage. President Wilson, with a tiny army at his disposal, wished to avoid involvement in the war. But the tide of opinion had turned decisively against Germany, and the loss of the *Lusitania* helped pave the way for the United States to join the war on the side of the Allies.

Even during the war, controversy raged over the cause of the violent explosions. Germany and her apologists asserted vociferously that the *Lusitania* had been carrying contraband explosives. The British and Americans replied with equal fervor that this was not true. The arguments continued after the war (indeed, they persist to this day). But the only way to verify the competing claims was to examine the wreckage, lying 295 feet down in the cold, turbid waters of the Irish Channel, amid powerful and treacherous bottom currents.

The first person known to have explored the wreck was an English diver, Jim Jarratt, who went down in 1935. Jarratt used a conventional helmeted outfit (he made his dive years before Cousteau invented SCUBA); with limited mobility, vision, and bottom time, he concluded that the ship lay on her port side. He also thought he saw holes in the forward part of the hull that looked as if they had been caused by an internal explosion.

In the same year, there were reports that the British Navy was blasting the wreck, possibly to test depth bombs or perhaps to open it up and recover evidence. This blasting supposedly continued for years. In the process, much damage was done to the remains of the ship.

In 1953, divers again visited the wreck and confirmed that it was lying on its starboard, not its port, side. This observation made sense since the torpedo had struck the ship on the starboard side. Then, in 1960, a young American diver named John Light claimed that he had dived on the *Lusitania*. His claim could not be confirmed, however, since there are over 7,000 confirmed wrecks in the area, and Light, who was SCUBA diving on ordinary air (a reckless venture at that depth), might well have been suffering from nitrogen narcosis and could easily have been confused. Light was also diving in a wet suit rather than a dry suit, and hypothermia could also have clouded his perceptions.

In the years that followed, Light, who was an ex-navy diver, and two other ex-navy men made over 100 dives to the wreck (at least according to their own account), but again it is not certain that they actually reached the *Lusitania*, for their reports were conflicting and fragmentary.

Light, obsessed with the *Lusitania*, somehow scraped together the money to buy her at auction—for a mere £1,000—from the British government insurance company that had acquired title to the wreck after paying Cunard's claim for the loss. This was in 1967.

Light became involved with Gregg Bemis and Bemis's partner, George Macomber, in the project to create a mobile saturation-diving system. He sold them part ownership in the wreck, on which they planned to test the system. According to Bemis, the saturation-diving system was 90 percent completed when the money ran out, and it had to be abandoned before it could be put to the test with humans. Light, who had pledged his ownership to his two partners in return for their financing, had to forfeit his claim. A few years later, Macomber sold his share to Bemis, leaving him the sole owner. (In 1982, the ubiquitous Oceaneering Inc. got permission from Bemis to dive on the wreck and came back with three of the four propellers and some other items. However, Oceaneering claimed that it had not made enough to cover its costs.)

By this time, Gregg Bemis, too, had become obsessed with the torpedoed liner and made it his goal to establish whether or not she had been carrying contraband. However, he did not visit the wreck personally until 1993, when he contracted with the famed Bob Ballard to explore the *Lusitania* using a small submersible and ROVs. Bemis visited the wreck in the little yellow *Delta* submersible, supplied by Delta Oceanographics, and inspected a section of the collapsing hull. Bemis and Ballard, two men possessed of strong and commanding personalities, did not get along, however. The expedition was funded by *National Geographic*, which produced a lavishly illustrated article in the April 1994 issue of its magazine. But it did not provide the information that Bemis was looking for.

In the *National Geographic* article (in which Ballard, tellingly, did not mention Bemis's name), Ballard stated that the most probable cause of the second explosion was coal dust and gave a long, well-reasoned explanation of his belief. (One fact in Ballard's account strongly supports his theory: eighteen boxes of percussion fuses remained intact; several were recovered in 1982.) Bemis remains unconvinced.

Bemis has returned to the *Lusitania* every summer since 1993; the divers he has hired speak well of him—but still have not turned up the evidence he seeks. The wreck is difficult to explore; it is draped in the usual snagged fishing nets and is filled with silt to within 6 feet of the top. A complicating factor in the search is a decision by an Irish court stating that, although Bemis holds title to the ship's hull and equipment, he does not own the cargo or any personal effects of the doomed passengers and crew. In addition, the Irish government has declared the *Lusitania* a national historic site, meaning it is a "hands off" area.

Bemis's interest in solving maritime mysteries got him involved in a highly controversial project to investigate the wreck of the huge Baltic ferry *Estonia*, jointly owned by Swedish, Finnish, and Estonian interests. The *Estonia* sank in a bad storm on September 28, 1994, with an appalling loss of life. Its sinking has spawned a rat's nest of speculation about the cause, from reasonable to wild and vilely bigoted conspiracy theories. Many people believe that the three Baltic governments are engaged in a long-term cover-up (see chapter 5), declaring the wreck site a sacred burial ground, not to be visited. In a suspicious move, the Swedish government attempted to bury the *Estonia* in concrete, officially to prevent its desecration by curious divers. (The attempt failed when the government, to save money, dumped tons of sand on the wreck to cut down the amount of expensive concrete needed for the job. Currents washed the sand away.)

Sweden, Finland, and Estonia, plus Denmark, Latvia, the U.K., and Russia, signed a treaty criminalizing any attempt to investigate the wreck. Bemis, who organized an expedition to the sunken *Estonia* in the summer of 2000, fell under this treaty. He was arrested *in absentia,* and, if he sets foot on the soil of Sweden, he will be jailed and tried. Undeterred, he is fighting the treaty in court. He firmly believes that the sinking was not an accident but deliberately caused. A stubborn seeker, indeed.

Quite a different type from the hard-driving Bemis is Clive Cussler, the best-selling author with books on the racks in bookstores all over the country. If an airport is big enough to have a store, the chances are that there will be a title by Clive Cussler in it.

Cussler was born in 1931 in Aurora, Illinois, the sickly son of a German army veteran of World War I. When he was eight, his family moved to southern California for his health, and young Clive spent his formative years there. A bright boy but a lazy student, he barely scraped his way through school. After a year in a two-year program at a community college, he joined the U.S. Air Force—the Korean conflict had broken out—and ended up in Hawaii, where he spent his free time skin-diving and spear-fishing with his air force buddies. After his military service, Cussler married and settled in Los Angeles. He progressed through a variety of jobs, from pumping gas to managing a filling station to selling encyclopedias door to door. Then he landed a job as advertising manager for an upscale supermarket. At last he had found his niche—for a while. By his own account, Cussler was a very successful ad man, but he began to grow tired of this line of work. When his wife got a night job in the local police department, he began writing adventure stories after their children had gone to bed. Recognizing that the market for tales of secret agents and detectives was overcrowded, he decided to make his protagonist a marine engineer. He sold a book and then, at his wife's urging, left his advertising career and went to work as a clerk in a dive shop. After a year, he returned to his lucrative career in advertising but fell victim to office politics and quit to become a full-time writer.

With the money he earned from his extremely popular novels, Cussler was able to indulge his childhood dreams of adventure and exploration. With friends, he founded the National Underwater & Marine Agency (NUMA), named for the fictional organization for which his protagonist and alter ego, Dirk Pitt, worked. A master of persuasion (think "ad man"), Cussler enlisted a distinguished board of directors, including Commander Don Walsh of the U.S. Navy, who made history in 1960 by accompanying the Swiss inventor Jacques Piccard to a depth of 35,800 feet in the South Pacific in the bathyscaphe *Trieste*; the late Dr. Harold Edgerton of MIT, who invented a pioneering stroboscope device for underwater photography; and the late Dr. Peter Throckmorton, one of the founders of underwater archaeology.

Cussler and his board decided to ignore the search for treasure lost at sea and concentrate instead on historic ships. One of his proudest ventures was locating and recovering the Confederate submarine *Hunley*, which sank the Union warship *Housatonic* at anchor in Charleston Harbor in the first successful torpedo attack in naval history. The *Hunley* and her crew met a poetically just fate when they were run down by another Union ship on its way to rescue the surviving crew of the *Housatonic*.

Cussler's first expedition, however, was to find the wreck of the *Bonhomme Richard*, famous as the flagship of John Paul Jones, which was sunk in a naval battle off the east coast of England in 1783. The expedition was a fiasco: the British archaeologist with whom Cussler worked was a fine historian but knew nothing about running an expedition, and they made every mistake possible. The British crew of the retired minesweeper that Cussler rented were surly and uncooperative. The wreck that the British archaeologist thought was the *Bonhomme Richard* turned out to be a World War I freighter torpedoed by the Germans. The only triumph that Cussler salvaged from the expensive outing was an anecdote. His rented minesweeper was moored offshore and bobbing in the waves. To reach it, the members of the expedition and their equipment had to take a ferry. Ship and ferry were both bobbing with wild lack of coordination. The ship's crew helped everyone on board—except Cussler, who was struggling with an awkward armload of paraphernalia. Soaked, Cussler finally clambered aboard, unassisted and ignored. Then he turned toward the crew, raised his hand, and told them that they had to save that hand. "Why should we bust our arse to save that hand, mate?" asked one of the more loutish members of the crew. "Because *this* is the hand that writes the checks," replied Cussler. The attitude of the crew improved remarkably, though not their incompetence and inefficiency.

A second hunt for the *Bonhomme Richard* also failed, despite Cussler's having a sturdy ship and a top-notch captain and crew. His team did discover a sunken Russian spy trawler; upon notifying the British authorities, they were shooed away from the scene.

But Cussler has scored a number of successes in addition to the CSS *Hunley*, most of them in relatively shallow water. One was the *Lexington*, a fast side-wheeler steamboat that sank in Long Island Sound on a stormy and bitterly cold January night in 1840. Cussler also found a number of Civil War warships, such as the USS *Cumberland*, rammed by the dreaded Confederate ironclad *Virginia* as she lay at anchor off the mouth of the James River in Virginia; the Confederate raider *Florida*; and several ironclad riverboats used in the western theater of the war. His strangest quest was to recover a steam locomotive that had crashed into Colorado's flood-swollen Kiowa Creek in 1880 when the bridge was washed out. After extensive searching with metal detectors and ground-penetrating radar, he found nothing. Eventually, a volunteer researcher discovered that the railroad company had waited to collect the insurance on the locomotive and then clandestinely recovered it under cover of night, disguised it, and fixed it up to go back into service.

In all, Cussler and NUMA have discovered nearly sixty historic wrecks. The work has involved hundreds of hours of research before an expedition even sets out and often more than one expedition. But Cussler is not a treasure hunter. His work is a labor of love and carries a sense of playful adventure. Cussler's philosophy is "If it ain't fun, it ain't worth doing."

Last of the trio is Greg Stemm, a Florida-based salvager-cum-archaeology buff. Stemm and his researchers comb naval records going back to the seventeenth century or earlier for reports of ship losses and then cross-check to see which have already been found. Unlike Bemis, whose goal is to solve controversial cases, or Cussler, whose aim is adventure, Stemm investigates wrecks for profit. His most recent success was the HMS *Sussex*, an English warship that sank in the Mediterranean near Gibraltar in 1694, loaded with gold and silver coins to the value of £1 million, worth many times that today. The eighty-gun frigate was bound for the eastern Mediterranean but on the way was to deliver the money to the Duke of Savoy, a small but strategically located state wedged between France and Italy. The aim was to buy the duke's allegiance to England in an ongoing war between England and France. However, the monetary inducement never reached the powerful and calculating duke, and the course of the war was changed.

Stemm became aware of the *Sussex*'s existence in 1995, when one of his researchers unearthed a diplomatic letter that referred to a small fortune lost in the sinking. He proceeded to hire more researchers to search naval archives in England, the Netherlands, France, and the U.S. They found a British document from November 1693 that said "A great sum of money is sending to Savoy." And the royal proceedings of December 12, 1693, just before the ship sailed, cited an order by the English king to send "a million [pounds] in money." That amount would be equal to 10 tons of gold or more than 100 tons of silver and would be worth an estimated $500 million today for the metal alone. In the form of coins, the find could conceivably fetch up to $4 billion if dealers and collectors were eager enough—but that is pure speculation. In 1694, when the money was not delivered, the duke, a famous general, declined to join the British, and the tedious and destructive Nine Years' War in Europe ground on.

Stemm and his partner, John Morris, almost had to wage a war of their own with the British government to obtain permission to salvage the cargo. Under maritime law, the salvor was entitled to a large percentage of the value of what was recovered—in some cases, as much as 90 percent. However, the *Sussex* was a government vessel, and governments are reluctant to relinquish their sovereign claims.

In the end, an agreement was worked out under which Stemm's company, Odyssey Marine Exploration, would split the proceeds with the government.

Before Stemm got even that far, he and his cohorts had painstakingly studied the currents and tides of the area where the *Sussex* went down, checked the logs of ships that had witnessed the disaster, and mapped the probable path taken by the admiral's body, which had washed ashore from the site of the sinking. Once permission was granted in 1998, Stemm and his team rented a French research ship, the *Minibex*, to explore the sea floor with side-scan sonar and ROVs.

Searching a 400-square-mile expanse of uneven sea floor with side-scan sonar, they found some thirty wrecks, ranging from a modern steel freighter to wrecks so ancient that nothing was left but their cargoes of ceramic amphorae, pointy-bottomed jars of all sizes that were used for storage and shipping. In fact, one of the first finds turned out to be the remains of an ancient Phoenician trading ship, lying almost 3,000 feet down and judged to date from the mid-fifth century B.C. (The date was estimated from the style of the amphorae.) The discovery caused great excitement among archaeologists, but it was not what Stemm was looking for. The *Minibex* also found a mound approximately the size of the missing warship. An ROV was sent down and transmitted back video images of long, narrow objects, possibly cannon, heavily encrusted with mineral deposits.

To rule out misidentification, Stemm and crew "mowed the lawn" in 50-mile tracks over their 400-square-mile search area. No other traces of cannon were found. In 2001, Stemm returned to the site. With the permission of the British Ministry of Defence and its Spanish counterpart (Spain, too, had a claim, for the wreck lay in Spanish waters), he recovered a small iron cannon and a cannonball. The cannon was made of iron rather than bronze, which pointed to a British provenance. Another clue was the absence of olive jars, which a Spanish warship would have carried. The evidence was enough to authorize further expeditions in 2002 and 2003.

Like Clive Cussler, Greg Stemm began his career in advertising and public relations. One of his first clients was the very popular comedian Bob Hope. On a whim in 1987, Stemm bought an 83-foot former shrimp boat that the Coast Guard had seized for running drugs. With one of his clients, John Morris, he founded a company called R. V. Seahawk to hunt for sunken ships. (Excitement over marine treasure-hunting had run high in Florida since the early 1960s, when Mel Fisher and others began harvesting gold and other valuables from sunken Spanish galleons in Florida's coastal waters.) The company went public as Seahawk Deep Ocean Technology and began selling shares to adventurous investors. Then disaster struck: the Securities and Exchange Commission began a seven-year investigation of the company for alleged securities violations. Finally cleared, Stemm and Morris founded a new company, which they called Odyssey Marine Expeditions.

Unlike Bemis and Cussler, who finance their investigations from their own fortunes, or the redoubtable Bob Ballard, who has but to submit a proposal to get funding from the government and a small army of institutions, Stemm has to scrounge up his money on his own for expeditions. Relying largely on the sale of shares in his expeditions to investors, he also sells merchandise, including a CD dictionary of colonial Spanish maritime terms. In 2000, he took tourists out to look for the *Sussex* at $2,250 a day per person.

In the summer of 2003, Stemm scored another coup: his company located the wreck of a nineteenth-century steamship that sank in a hurricane off the coast of Georgia in 1865. The ship, the SS *Republic*, was carrying a cargo of an estimated 20,000 twenty-dollar gold coins. Experts believe that these coins might fetch as much as $150 million from avid collectors. A freelance researcher put Stemm on the trail of the *Republic*; the search eventually covered an area of more than 1,500 square miles (1,000 of them searched in the last two years) and turned up twenty-four "targets," of which the last one was the payoff. To put rival salvors off the scent, Stemm gave his project the code name *Bavaria*. He used the latest in side-scan sonar and a towed magnetometer to locate the target and an ROV for the close-up investigation of the wreck, which lay a bit over 1,600 feet down. To clinch his claim under Admiralty law, his team brought back a bottle from the wreck as evidence.

The resourceful Stemm has had many a conflict with archaeologists, who have traditionally regarded salvors as their enemies.

Salvors, the archaeologists complain, care only for the value of the treasures they bring up and play havoc with any attempts to reconstruct the history of the wrecked vessel. Furthermore, the objects that the salvors recover typically disappear into private collections, where the public cannot see them and archaeologists cannot study them. This was all too often true in the past, though things are changing. For their part, the salvors say that without their efforts artifacts of historic value would simply stay on the sea floor until they are corroded away, and no one would ever see or study them.

Stemm is one of a new breed of salvors since he himself is part amateur archaeologist. A member of Mensa, a prestigious intellectual society, he has presented numerous papers and has participated in international panels. He always includes an archaeologist on his expedition teams and strives for cooperation between archaeologists and salvors.

Stemm foresees a great boom in the search for wrecks in years to come since he estimates there are over three million wrecks on the floors of the world's seas, and ROV technology that will enable seekers to reach them is improving rapidly.

8 Tragedy in the Sky: Air and Space Disasters

Although most undersea investigations deal with sunken ships, many airplanes crash into the sea under mysterious circumstances. And one U.S. spacecraft has met the same end. Information on the causes of these crashes, and even the investigations themselves, are hard to come by, since firms like Oceaneering observe a policy of strict confidentiality for their clients. This confidentiality is especially important to airlines, which like to concentrate attention on the services they allegedly provide in order to promote a positive image and attract customers. They do not want travelers to think about air disasters—that is embarrassing and bad for their reputations. Some airlines are owned by governments, and governments are even more secretive about the details of accidents and tragedies. In the case of military aircraft, governments are fearful that strategic secrets may be revealed if particulars about crashes are made public. Nonetheless, there are a few cases of air and space disasters for which information about the deep-sea recovery efforts is available.

* * *

On July 18, 1996, American television viewers, radio listeners, and newspaper readers were shocked to learn that a giant Boeing 747 jetliner, TWA Flight 800, had burst into flames and crashed into the Atlantic Ocean within sight of Long Island. The flight had just left

JFK International Airport for Paris with 230 people on board when the tragedy occurred. The flaming wreckage plunged into the sea in relatively shallow water, only 10 miles south of the town of Moriches, Long Island. The disaster was seen by many people on land; in fact, the Coast Guard duty officer learned of it from people phoning in to report large balls of flame falling into the sea. The Coast Guarder called the Air National Guard to see whether it had been practicing dropping flares. The Air National Guard checked with one of its own planes that was flying in the area. The pilot reported that there appeared to be a large commercial jetliner down, with many bodies floating in the water. The Coast Guard had a station nearby, and within twenty minutes it had two boats on the scene. Boats and aircraft—*surface* and *air assets* in the official jargon—from other government agencies quickly joined the search for survivors. They found none, although they recovered 100 bodies by 3 A.M. on July 19. In all, 216 bodies were eventually recovered.

The first objective of the search was, of course, to rescue anyone who had survived the crash. The second was to recover as much of the wreckage as possible in order to ascertain the cause of this traumatic event. Still another mission was to handle the media, which soon numbered representatives from over 400 organizations, backed up by over 25 TV satellite trucks.

The plane and its luckless passengers and crew fell into relatively shallow water—ranging from 115 to 130 feet deep—which placed it within the reach of divers. But the strong currents and poor visibility in the turbid water—12 to 15 feet on a good day and zero on a bad day—made their work extremely difficult. The navy divers were not entirely blind, however, for they were armed with handheld sonars that let them "see" the bottom and what was on it.

A total of 240 navy divers (not all at the same time), aided by divers from the FBI, the New York State Police, the Suffolk County Police, and the New York City Police, were involved in the search. According to navy data, divers made 3,167 SCUBA dives and 677 dives supplied with air from a surface vessel and spent a total of 1,689 hours underwater, not counting decompression time. The divers had special suits to protect them from the chilly (50-degree) water but still could work for only an hour at a time. Two mini-ROVs, supplied by Oceaneering, supplemented the search.

The navy began by staking out a 75-square-mile search area, which was soon reduced to 25 square miles when the debris field was roughly mapped out. It took three days to locate the debris field as the search ships slowly "mowed the lawn," but patient work was rewarded by the discovery of a 45-foot piece of one of TWA 800's wings. Within a week, divers had found and recovered the grand prize of any airplane crash investigation: the "black boxes"—the cockpit voice recorder and the flight data recorder—in 150 feet of water. Their search was aided by a towed device called a pinger locator system (PLS), which detects the pings made by the black boxes after a crash. Global positioning satellite (GPS) systems on the surface vessels allowed the divers to establish the precise locations of their finds for later recovery.

Working around the clock, divers recovered over 50 percent of the wreckage within the first month. Truckload after truckload, the recovered fragments were taken to a Grumman Aircraft hangar at Calverton, Long Island, where specialists painstakingly reassembled them. The divers continued their work through the summer and fall of 1996; the navy then hired four scallop trawlers to recover the remainder of the debris, which was mostly embedded in the sea-floor sediments. The trawlers worked through the winter and continued until late April 1997, when the investigators decided that not enough pieces were being recovered, and those that were dredged up were not of great significance. In all, about 95 percent of the wreckage was recovered.

Theories about the cause of the crash abounded, some truly irrational. One New Age enthusiast was convinced that the plane was targeted by a kind of death ray controlled by a sinister international organization. Anti-Semites ranted that the crash was caused by Israel to frighten American airlines into using Israeli security guards. Leftists hissed that the CIA had shot the plane down. Some government haters were convinced that the navy had shot down TWA Flight 800 because it was trespassing on a training area. Less bizarre was the theory propounded by a Harvard professor of aesthetics and the general theory of value. The professor, who apparently has a strong interest in science as well as in aesthetics, postulated that the plane, which was in fact flying through a naval training area, had accidentally been affected by electromagnetic interference from an experimental device the navy was testing.

But three major theories prevailed. The first was that the aircraft suffered a mechanical failure of some kind. This is a frequent cause of airplane accidents, and TWA 800 was an old plane, built in 1971. A second theory was that the navy had sent a missile into the plane by accident during a training exercise. This theory was plausible for several reasons: training accidents and accidents by "friendly fire" occur all too frequently, in all branches of the armed services and in all countries, although we don't hear about those that happen in countries other than our own. This naval error theory gained great popularity when it was taken up by Pierre Salinger, the former press secretary of President Kennedy and a respected television commentator. Right-wing ideologues seized upon this theory as evidence of a massive government cover-up (the accident occurred during the presidency of William Jefferson Clinton, a Democrat).[26] The third theory was that Muslim terrorists had placed a time bomb on the plane—a believable speculation in the light of the destruction of Pan Am Flight 103 over Lockerbie, Scotland, by a terrorist bomb and the 1991 attempt by Muslim extremists to blow up the World Trade Center.

After spending months analyzing the wreckage and sonar images of the debris field, the National Transportation Safety Board (NTSB) came up with a much less dramatic conclusion: the plane had been destroyed by an explosion in the center-wing fuel tank, which was nearly empty except for vapors. The explosion was caused, said the NTSB, by a short circuit in electrical wiring above the tank, which had sent a jolt of current to wires inside the tank. Examination revealed that the wires had been damaged by repair work done at various times on the plane, and there was evidence of metal drill shavings in

[26] One of Pierre Salinger's co-authors, Ian Goddard, later confessed that he had promoted the conspiracy theory to "give the government a black eye by any means that looked opportune." The crash of TWA 800 was merely a vehicle for his larger agenda, he said.

the hollow places inside the aircraft structure. This debris could have chafed the wiring and worn away insulation, permitting a fatal spark.

* * *

Another noteworthy airplane disaster was the crash of Swissair Flight 111 on September 2, 1998. It was a quiet, moonless evening when the jetliner took off from JFK International Airport on Long Island, bound for Zürich, Switzerland, with 227 passengers and crew members on board. Around 10:30 P.M. Atlantic time, the pilot issued a *Pan-Pan-Pan* call (international aircraft code for a problem that has not yet turned into an emergency). "We have, uh, smoke in the cockpit," he told the Canadian air traffic controller at Moncton, New Brunswick. "Uh, request immediate return, uh, to a convenient place, I guess, uh, Boston." Then muffled voices from the cockpit suggested that pilot and co-pilot were putting on their oxygen masks.

The controller at Moncton asked if the pilot would prefer to land at Halifax, Nova Scotia, and the Swissair pilot agreed. About six minutes later, the pilot told a Halifax controller that he needed to dump fuel (which is standard procedure before making an emergency landing), and he wanted to do this over water. About three minutes later, just after the Halifax air traffic controller had cleared the plane to start dumping, its radio went silent. Six minutes later Swissair Flight 111 struck the water.

As the Halifax controllers monitored Flight 111's progress on radar, they lost it seven miles south of a coastal village called Peggy's Cove. That position gave the last known location of the plane. Before dawn, the Canadians had dispatched a fixed-wing Hercules plane, two helicopters, and two navy ships to search the area. They were joined by the local fishing vessels. But visibility was so poor that it was almost impossible to see anything. Everyone knew, though, that something terrible had happened.

In the nearby town of Blandford, the proprietor of a bed and breakfast told of how she and her husband had been lying in bed when, suddenly, the whole house began to rumble. She heard a sound like the noise of a helicopter or something else "tearing the sky apart." A few seconds later, she heard a thud as the plane struck the water. She reported no explosion. Another Blandford resident, a librarian, heard the plane as it passed overhead but thought it was nothing unusual, since aircraft flew over her house all the time. Then came what she called an explosion that was not really a sound but more like a big vibration.

It was uncertain at first who should investigate the crash. The Canadians, because the crash occurred in Canadian territorial waters? The Swiss, because the plane was registered in Switzerland? The United States, because the plane was built in the U.S. and had departed from an American airport? As far as the U.S. Navy was concerned, the question was moot, since the United States was treaty-bound to render assistance. In fact, the United States did send assistance in the form of thirty-two divers and the USS *Grapple*, a salvage-and-rescue ship capable of raising 300 tons from the seabed. The U.S. Navy also provided state-of-the-art equipment: a synthetic-aperture sonar and a laser-line scanner. This type of scanner, towed just above the sea floor, uses a blue-green laser light to obtain video images of startling clarity. According to the navy, it can display details of objects that are only 0.5 inches across and cover an area up to five times that of a conventional underwater camera while still retaining clarity. (Normally, the wider the focus, the blurrier the picture.) Both the synthetic-aperture sonar and the laser-line scanner were part of a newly developed system for locating mines; for the Swissair crash, a military tool was turned to a civilian purpose.

The *Grapple* also served as the base of operations for the ROV *Deep Drone.* Fairly good-sized for an ROV, *Deep Drone* has a depth capability of 7,200 feet and is equipped with two manipulators, target-locating sonar, a 35-millimeter still camera, one black-and-white camera, and one color video camera. It measures 9 feet 3 inches long, 4 feet 7 inches wide, and 6 feet 2 inches high, and it can carry a payload of 300 pounds—clearly, a vessel designed for heavy work.

Despite the depth of the water (270 feet) and strong currents, the investigators decided to use divers to help recover the fragmented bodies of the victims and search for the two vital black boxes. One of these boxes is the flight data recorder, or FDR, which automatically records such pertinent data as altitude, airspeed, heading, attitude of the wing flaps and rudder, the on or off position of switches, the moment when autopilot is disengaged—as many as 1,000 parameters if so programmed, including whether the smoke alarms in the lavatories are on. All of this information is recorded on chips or magnetic tape, to be deciphered later by specialists. The data from the FDR allow investigators to do a computer simulation of the flight up to the moment when the recorder stops working.

The other box—the cockpit voice recorder, or CVR—records not only the pilots' voices but all noise of any kind in the cockpit, including the sound of the electric motors that adjust the pilots' seats and the clunk of the cockpit door opening and closing. In newer airplanes, this information is stored in digital form in a solid-state device that holds up to two hours' worth of sound. Older planes record it on a loop of magnetic tape that holds only thirty minutes' worth of data before it starts rerecording over the previous material, limiting the knowledge available to crash investigators. Swissair 111 had the thirty-minute tape.

Black boxes are actually colored bright orange to aid in their visual location (if the plane crashes on land) and carry a pinger that can be picked up by sonar if the plane crashes into a water body. They are called *black* because the material they contain is confidential. Pilots do not want any of their indiscreet or personal remarks to be revealed when the device is played back. There is always a risk of disciplinary action to punish remarks that are made under stress.

Black boxes are typically made of heavy-gauge stainless steel or titanium. Inside this shell is a layer of heat insulation, then a layer of a waxlike insulation around the memory unit. Black boxes are rigorously tested for impact shock resistance. One method fires them from a 40-foot cannon at 355 miles per hour into an aluminum "cushion" that contains sensors and gauges to record the impact as the box progresses through it. To test resistance to penetration, engineers drop a 500-pound weight with a protruding steel rod onto the box at its weakest point. Crush resistance is tested by subjecting the box to a force of 5,000 pounds for five minutes. Afterward, the memory unit inside the box is fished out and examined to establish how well it fared.

The FDR is typically located in the tail of the plane because that is the least likely part of the aircraft to be damaged in a crash. The CVR is, naturally, located in the cockpit. Despite their strength, black boxes are occasionally destroyed in a crash, but this happens very seldom.

The divers assigned to search for Swissair Flight 111's black boxes, both Canadian and American, were outfitted with the old-style helmet suits. Some had the amenity of a hot-water hose supplied from their support vessel, which allowed them to spend more time on the bottom. Otherwise, they could not work for more than thirty minutes at a time, including descent time. They had to operate with extreme caution to avoid getting their air hoses or suits torn on the jagged edges of the wreckage. Ascent for these divers required an hour or more, in slow stages. After the ascent, they had to go into a decompression chamber on deck and spend two hours breathing oxygen. Because the decompression chamber could accommodate only two divers at a time, the pace of work was slowed.

Nevertheless, on Sunday, September 6, after only four days of concentrated effort, the flight data recorder was retrieved from the ocean floor. The voice recorder was brought up six days later. Information leaked to the press revealed that the pilot and co-pilot had spent the last moments of their lives arguing over whether they should disregard protocol and head directly for Halifax or go by the book and dump their fuel first. The co-pilot urged the pilot to head for Halifax, while the pilot insisted that the rules must be followed, even as the plane's electrical system was deteriorating. The revelation of this cockpit exchange was felt to be a breach of solemn trust and caused outrage in the aviation community.

Meanwhile, the sad job of bringing up the wreckage continued. Bit by bit, the jagged pieces were raised and pieced together. The wiring—the plane had 150 miles of it in all—was found to be severely scorched in the forward area. There was evidence also of arcing where the insulation on the wires was lacking. This discovery reinforced the suspicion of the Canadian Transportation Safety Board (TSB) that electrical failure had doomed the aircraft, since both the flight data recorder and the cockpit voice recorder had ceased working at the same time, six minutes before the actual crash. Further investigation revealed that the wiring was insulated with a type of plastic called Kapton, which is technically an aromatic polyimide film.

Kapton had already been effectively banned in U.S. military aircraft but not in civilian craft. The problem with Kapton was that it deteriorated too quickly under use and cracked. The uninsulated spots on the wires could then cause short circuits or, worse yet, spew out an electric arc, risking a fire. And the melting insulation could form a carbon deposit, like the electrode in a dry-cell battery, which could act as a conductor and escalate the short-circuiting into a cascade of things going wrong. It was later suggested by aviation experts that the straw that broke the proverbial camel's back was the crew's turning on the in-flight entertainment system. This voracious consumer of electrical power apparently caused the wires to overheat and short out. The power failed. Without power, the pilot and co-pilot could not work the aircraft's controls. The rest is history.

Recovering the wreckage continued through the fall and into early winter, with a scallop dragger replacing the divers. Much of the material showed signs of "heat distress," as the TSB delicately put it. There were small pieces of melted aluminum from the ceiling just aft of the cockpit door, melted copper from the wiring, and marks of arcing. A few drops of melted plastic were found embedded in the sheepskin cover of the observer's seat, just behind the pilot.

Twenty-seven months after the accident, the TSB had not completed its analysis of the cause, although by December 1998 it was obvious that fire was the culprit. Two million pieces of wreckage had been recovered, studied, and documented. The results were a series of recommendations from the TSB to improve fire safety on board aircraft, including installing built-in fire-suppression systems; training crews in fire-fighting and detection; introducing a simplified checklist for emergencies (the regular checklist took thirty minutes to complete); and, not least, mandating crews to override protocol and land their planes as quickly as possible after smoke is detected. The TSB's final report, issued in March 2003, confirmed the original suspicion of fire as the cause of the crash, and the board added nine additional aircraft safety recommendations.

* * *

EgyptAir Flight 990 was scheduled for a long trip when it took off from Los Angeles International Airport on October 30, 1999, bound for Cairo with a stopover at New York's JFK International Airport. At JFK, it changed the flight crew, taking on two captains and two first officers. The double set of pilots was required by international regulations for the ten-hour flight from New York to Cairo. In addition, there were 10 flight attendants and 203 passengers, 100 of them Americans. Fifty-four of the Americans were heading for a two-week tour on the Nile. Also aboard were thirty-three Egyptian military officers and a group of off-duty EgyptAir pilots deadheading back to Cairo. One of them was the airline's chief pilot; some were junior pilots working their way up the promotion ladder.

The plane left JFK uneventfully at 1:20 A.M. on October 31, the passengers presumably settling drowsily into their seats for the long flight ahead. The captain for this leg of the flight was Ahmad al-Habashi, aged fifty-seven, described as a portly, jovial man. He had a bad back but was otherwise healthy. The relief first officer, or co-pilot, was Gameel al-Batouti, aged just under sixty, described as a big, friendly man getting ready for his retirement. Before joining EgyptAir, Batouti had served as a flight instructor at Egypt's aviation academy. There was also a junior co-pilot, Adel Hatem.

Half an hour out from JFK, at an altitude of 33,000 feet and 60 miles south of Nantucket Island, Massachusetts, EgyptAir 990 suddenly plunged downward and then disappeared from the radar screen. The story of this mysterious accident would eventually be deduced from the evidence contained in the flight data recorder and the cockpit voice recorder. But first they had to be found.

Naval charts and sonar readings showed that the water in the area averaged some 270 feet in depth; recovering the wreckage would be a major job. Who should undertake it? By rights, it should have been the Egyptian government, since the crash occurred in international waters and the airplane belonged to Egypt. However, the Egyptians handed the task over to the National Transportation Safety Board and to the U.S. Navy, which called upon Oceaneering to do the search-and-recovery work.

Within hours after the crash, four ships and eleven aircraft were searching the area where the plane had vanished from the radar screen. The first sonar readings over a 36-square-mile area revealed that the debris was concentrated in two main fields. The western field, which was smaller, contained the plane's left engine. Most of the plane lay some 1,200 feet to the east. The fact that the debris was concentrated rather than widely scattered indicated that the plane had not exploded or broken up in the air.

The navy sent the rescue ship *Grapple*, which had served well in the earlier two incidents described in this chapter, plus an ocean-going tug equipped with a pinger locator and a mine-hunting vessel. Two ROVs were involved in the search: the navy's *Deep Drone* and Oceaneering's *Magnum*. The plan of action was for the ROVs, using grapple hooks, to drag the debris into piles; helmeted divers would then load it into salvage baskets that would be hoisted aboard a surface vessel. Some of the debris was gathered with a crane and a clamshell bucket. On the support vessel, front-end loaders shoveled it into containers, which were then taken ashore. There the containers were rinsed twice and moved to a hangar at Quonset Point, Rhode Island. Tipped onto their sides, they were emptied by a human crew armed with shovels and rakes. A new set of front-end loaders spread the fragments of metal and plastic across the hangar floor to dry. While the debris was being spread out, FBI agents and NTSB investigators inspected each item for signs of fire or explosion damage, such as blackening or massive distortion. They found none.

The flight data recorder was recovered on November 2 by the reliable *Deep Drone*, which also brought up the cockpit voice recorder on November 13. The FDR showed the airplane flying normally for the first half hour until a strange sequence of events occurred. The autopilot disconnected. The throttles of both engines were yanked back to minimum idle speed, and the elevator flaps were moved to the "down" position. The plane dropped precipitously from 33,000 feet to 16,000 feet, nearly reaching the speed of sound, far beyond its safety margin. The right and left elevators began to move in opposite directions, the left one up and the right one down—a maneuver termed a *split* in aviation jargon. A mere second or two later, the engines were shut off. Then both throttle handles were moved forward, as if to gain power. Then the speed brake flaps were deployed. The plane abruptly reversed its dive and climbed to about 25,000 feet before plunging again. At 1:51:15 A.M., it went into a second dive. It hit the ocean a little more than sixty seconds later.

A delegation of Egyptian experts was flown quickly to the United States to take part in the investigation. They relayed the data from the FDR to Cairo, where a high-ranking official announced that the FDR data were inconclusive and that the accident could have been caused only by a bomb in the cockpit or the lavatory right behind it. The official did not know that the cockpit voice recorder had been recovered the night before and given to the NTSB. The CVR's tape was cleaned and processed, and a small group that included an Arabic translator listened to it at NTSB headquarters. The tape seemed to explain a good many things that appeared on the FDR.

After takeoff, the plane was put on autopilot, and Captain Habashi relaxed with the junior co-pilot, Hatem, asking him for the chart of the Atlantic. A flight attendant asked Habashi and Hatem if they would like some coffee. Hatem couldn't find the chart. "It seems that Samir left it, sir," he told Habashi, who was coughing. The CVR did not indicate whether the chart was ever found, but Hatem expressed his thanks to Habashi, presumably for not reprimanding him.

The chief pilot came into the cockpit, chatted for a bit, and left. Then, about twenty minutes after takeoff, Gameel al-Batouti came in. He and Habashi were old friends; Habashi called him "Jimmy." Batouti urged the junior co-pilot to go back into the cabin and get some rest; Batouti would stand in for him even though it was not yet time for the crew to change over as per schedule. The junior pilot wasn't ready to turn in, and it took considerable urging from Batouti to persuade him to surrender his seat.

Habashi and Batouti had a satisfying session complaining about the younger pilots, whom they agreed were arrogant, cliquish, and doing their best to undermine the two older men—a situation not at all unlikely in the highly politicized, state-owned EgyptAir. The cockpit door opened again, and one of the young off-duty pilots came in, dressed in civvies. "What's with you? Why did you get dressed all in red like that?" asked Habashi disapprovingly. A flight attendant brought dinner for Batouti, who pronounced it excellent but declined a second helping.

At 1:48 A.M., Captain Habashi excused himself to go to the toilet. "Go ahead, please," replied Batouti. "Before it gets crowded ... while they are eating, and I'll be back to you," Habashi elaborated. There were sounds of an electric seat motor operating and the cockpit door opening, then a thunk and a clink. Batouti was presumably left alone in the cockpit while his friend relieved himself. One of the cockpit microphones picked up an unintelligible three-syllable utterance that might have been "hydraulic" or "control it." Then Batouti, in a low, calm voice, uttered three words in Arabic: "Tawwakalt ala Allah." This phrase was initially translated as "I put my fate in the hands of God," but the Egyptian delegation objected that it really meant "I rely on God."

According to some accounts, this phrase, widely used in the Arab-speaking world, is used to ward off trouble for a whole range of incidents, from serious to trivial, much as the Western world says "Knock on wood." But other sources say that it is used after someone has arrived at a serious decision and wants God's support.

The most probable scenario is that Batouti then switched off the autopilot, cut the engines, and put the elevators into the down position. There were a number of loud thumps. Batouti repeated "I rely on God" eight more times before Habashi rushed in, alarmed. "What's happening, what's happening?" he cried.

"I rely on God" was Batouti's only reply.

"What's happening?" Habashi asked again.

Silence from the co-pilot.

At this point, Habashi apparently reached his seat and grabbed his control yoke, pulling it sharply backward to arrest the plane's descent.

"What's happening, Gameel? What's happening?" he yelled again.

Then Batouti apparently killed both engines. The plane was now powerless. Habashi shoved the throttles forward in a vain attempt to restart the engines. Meanwhile, Batouti was pushing hard on his control yoke, keeping the plane's nose down.

"What is this? What is this? Did you shut the engines?" Habashi asked. Then he shouted, "Get away with the engines! Shut the engines!"

"It's shut," replied Batouti.

"Pull ... pull with me ... pull with me ... pull with me," Habashi pleaded.

Batouti pushed instead. The plane went into a spin as it sped downward, the left engine came off, and the fuselage began to break up from the intolerable stresses placed on it.

To the NTSB, the obvious conclusion was that Batouti, for whatever reason, had deliberately brought the plane down. Suicide was the suspected motive. But when the NTSB's chairman, James Hall, held his first news conference, he was scrupulously neutral. Castigating the media for reporting the leaked information on the contents of the CVR, he sedulously avoided using the words *Arab* or *Muslim* or *suicide.* He did not even mention Batouti's name. Instead, he said that the accident "might, and I emphasize *might,* be the result of a deliberate act." He concluded with a soothing statement: "The truth is what both the American people and the Egyptian people seek."

The Egyptians, however, reacted with fury. They cried that the NTSB's preliminary conclusion was a canard: Batouti was being scapegoated because he was Egyptian and the plane was Egyptian. They also pointed out that Bernard Loeb, the NTSB's chief aviation investigator, was Jewish. The Egyptians smelled a plot by Israel or the CIA, or both, to discredit Arabs and all Islam. They suggested at first that a bomb had been placed in the tail of the plane by Islamic fundamentalists, who were waging a ruthless guerrilla war against the secular Egyptian government. When neither the NTSB nor the Egyptian team could find signs of bomb damage, the Egyptians insisted that the plane had suffered a mechanical failure and that the NTSB's scenario was the expression of a preconceived plan. As for the suicide theory, both the delegation and media all over the Muslim world said that it was impossible: suicide was contrary to Islam. (Suicide bombers are considered martyrs, not sinners, because they die fighting the "infidel.")

On the face of it, Batouti did not seem a likely candidate for suicide. He was married, with four grown children and a ten-year-old daughter who was being treated for lupus in Los Angeles. He had bought a tire in New Jersey to take home, and he also had some free samples of Viagra to give his friends. Nor did he fit the mold of the Islamic terrorist. He was only moderately religious, drank a bit (alcohol is forbidden to Muslims), and had no known connections to terrorist organizations. (On the other hand, Al Qaeda's members do not publicize their connection to that organization.)

The Egyptian delegation produced many arguments, some of them strong, to exonerate Batouti and EgyptAir. But a puzzling fact remains unanswered. Among the things that experts can interpret from the CVR are the tone of voice, emphasis, articulation, timing, breathing rate, and choice of vocabulary of the personnel in the cockpit. And while Habashi's voice showed agitation and alarm, Batouti's soft declarations of faith remained calm and serene until the end.

* * *

After a severe frost the night before, the morning air was still chilly on January 28, 1986, as the crew of the space shuttle *Challenger* prepared for takeoff at Cape Canaveral, Florida. The shuttle was actually a four-piece assembly that included the orbiter, the glider-like craft that carried the astronauts and scientific equipment; a gigantic, 164-foot external fuel tank filled with liquid hydrogen and oxygen; plus twin solid-fuel rocket boosters. All of these components were held together by specially designed bolts, which would be severed by explosive charges at the proper moment.

The boosters would give the shuttle the additional lift it needed to get off the launching pad and drop off when it reached 150,000 feet, or 28.4 miles, of altitude. The external tank, powering the orbiter's main engines, would drop off when the shuttle reached its orbit level, which varies between 115 and 250 miles above Earth, depending on the mission. The boosters would descend by parachute into the Atlantic Ocean, hopefully at a predetermined point, where they would float until retrieved. The external tank would mostly burn up in the atmosphere; the rest would drop into an empty region of the Pacific Ocean or over thinly populated areas of land. The orbiter itself, with the aid of smaller rocket thrusters, would maneuver itself into and out of orbit, adjusting its attitude with remarkable sensitivity. It could even dock with other craft in space. Once the mission was completed, the orbiter would return to Earth, slowing down as it passed through the atmosphere, and eventually glide to a landing.

Challenger's mission was to ferry a tracking data relay satellite into orbit and to launch a free-flying module that would observe the tail and coma of Halley's Comet. While this scientific portion of the mission was fairly ordinary by now, this particular flight had a unique feature: one of the astronauts was Sharon Christa McAuliffe, a high school teacher from New Hampshire. Mrs. McAuliffe had been chosen from more than 11,000 applicants to initiate a program called Teachers in Space. This program was intended to demonstrate the Reagan regime's commitment to education, and a great deal of political capital was riding on it.

The *Challenger* had made nine round trips successfully, and it was expected that this one would be successful, too. But, seventy-three seconds after liftoff, viewers were startled to see a puff of smoke and flame erupt from the side of the fuel tank. Then, as they watched in horror, the shuttle tipped onto its side and exploded, falling into the Atlantic and carrying its seven passengers to their deaths. Debris large and small rained down almost from the beach to 40 miles offshore.

(According to an article in the *Miami Herald*, what occurred was technically not an explosion but a fire and ensuing violent breakup caused by aerodynamic stresses as the shuttle assemblage hurtled through the air at many thousands of feet per second. If there had been a true explosion, the paper pointed out, it would have been so violent that there would have been little chance of recovering much identifiable debris.)

For the National Aeronautics and Space Administration (NASA), it was imperative to recover the remains of the astronauts to assuage public opinion. From the viewpoint of future shuttle launchings, it was even more important to recover the remains of the shuttle so that NASA's engineers and scientists could learn what had gone wrong. It would be an extremely difficult task, since the debris was scattered over 450 square miles, in water that ranged from 50 to

1,200 feet deep. There were strong currents—up to 7 knots—and unknown bottom topography to contend with. The job went to Oceaneering as the prime contractor. Surface vessels were supplied by Tracor Marine, and the Harbor Branch Oceanographic Institution (HBOI) provided two manned submersibles, *Johnson Sea-Link I* and *II*, each with a depth capability of 3,000 feet.

The search took seven months, and at its peak Oceaneering was riding herd on eight surface vessels, manned submersibles, ROVs, and teams of divers. Within three days after the accident, Oceaneering had its first teams on the site and had begun mapping the sea floor and locating wreckage with side-scan sonar.

Frantically nervous about bad publicity, NASA insisted that every bit of debris recovered had to be brought in under a tarpaulin. As the captain of one of the support vessels commented, there was no part of this mission that was any fun. The weather was terrible, the seas were "horrendous," and the 7-knot bottom currents were something that no trained diver would go near if it were not required by the job.

The search-and-recovery teams had to contend with nonhazardous problems as well. The highly sensitive sonar picked up not only debris from the shuttle but also conch and clam shells, fish beds, and coral heads. And the search area lay under a major shipping lane, which meant that oil drums, coils of wire, refrigerators, and a good deal of other junk that had either been dumped or washed overboard from freighter decks in storms littered the bottom. There was debris from earlier shuttle flights, and there were the remains of wrecked airplanes and ships. The Rogers Report, commissioned by the president, stated that, of 490 "contacts" classified, 408 did not come from the *Challenger*.

The *Miami Herald* article gave a grim perspective on the recovery effort. The first big find was made on February 8, when a sonar towfish picked up a large object on the bottom in 90 to 100 feet of water, about 15 miles northeast of the launch site. It turned out to be the crew compartment. The NASA flight controllers were sure that the crew compartment would have disintegrated on impact, for water behaves like concrete when an object strikes it at high speed. Instead, it had been held together by the thousands of feet of wiring that surrounded it. Inside were the bodies of the seven unlucky astronauts.

For political reasons, recovering the bodies of crash victims is usually given a very high priority. However, in this case, NASA's most urgent goal was to recover the failed parts of the shuttle in order to analyze them. Not until March 7 were divers sent to investigate "Target 67," the code name for the crew compartment. It must have taken a strong stomach to conduct this part of the investigation. One of the divers saw a pair of white legs protruding from the debris, waving gently in the current. He recoiled in horror, only to find that it was an empty space suit, taken on the mission for space walks.

The real bodies were another story. They had been torn apart by the force of the impact. One find was a space helmet containing the scalp and ears of its wearer—it was no wonder that NASA wanted to keep such details hushed up. The bodies had been partly eaten by crabs and other sea-floor scavengers; what was left of them had become almost gelatinous and ready to disintegrate. Some of the flesh had acquired a sort of soapy or waxy appearance due to biochemical reactions with the seawater.

The first three bodies—including the scientist Dr. Judith Resnik and the "Teacher in Space" Christa McAuliffe—were brought up on March 8 and placed in black plastic body bags aboard the navy auxiliary rescue-and-salvage ship *Preserver*. They were taken ashore, concealed in 30-gallon garbage cans, and taken in a navy pickup truck by night to a hangar at nearby Patrick Airbase—hardly appropriate treatment for those whom the president himself had hailed as heroes. The local medical examiners, who were legally responsible for conducting autopsies, were prevented from doing so. Instead, a doctor selected by NASA examined the pathetic remains and announced that the cause of death was impossible to determine. (NASA had previously claimed that the bodies had been all but vaporized.) In the interest of damage control, NASA said that it would make no statement since it did not want to upset the families of the astronauts.

About three weeks after the accident, HBOI was called in with its two submersibles. It was *Johnson Sea-Link II* that found a piece of the booster rocket with a defective O-ring. Designed for oceanographic investigations, the little sub has limited lifting capability, so the heavy relic was hoisted up by a winch. The two submersibles worked on through May, mostly identifying and helping to recover pieces of debris from deep water. When a piece of debris was located, they would attach a grapple to the chunk of metal, and then a winch on the support vessel would bring it up.

The flight recorders and other critical objects were treated with the greatest of care when recovered. As soon as they were brought up, they were placed in a bucket of iced seawater, iced fresh water, or fresh water at ambient temperature according to their nature. This protected them from further deterioration by exposure to the air. The process was something like the first stages of conserving archaeological artifacts brought up after centuries of immersion under the sea.

The *Miami Herald* reporter noted that the space shuttle carried no emergency locating transmitter, or ELT, the aircraft equivalent of a ship's EPIRB. The official reason for this omission was that the device weighed 10 pounds, and every ounce of weight is critical on space flights. However, in the crew cabin were "700 embroidered mission patches, more than 1,600 flags of various sizes, countries, and states, a video disk, an assortment of medallions, a deflated soccer ball, the town seal of Framingham, Mass., 47 copies of the U.S. Constitution, and patches, pins, and other things for organizations that wished to have something that had been in orbit aboard the shuttle."

Apparently, NASA was so confident that nothing could go wrong with a shuttle mission and so convinced that no one would survive a crash that it thought a locating transmitter was superfluous. For saving lives, it probably was, but it could have guided searchers to the crash site and saved valuable time and money.

Examination of the critical booster section and other wreckage revealed that an O-ring that was supposed to seal a joint between sections of the rocket booster had failed. The hot gases, jetting out at high pressure, had enlarged the small hole and burned through the wall of the external fuel tank. At the altitude that the shuttle had reached, the oxygen was so thin that this failure might not have been fatal in itself—but the flame also burned through one of the supports that separated the booster from the big fuel tank. The booster, which at that point probably weighed at least 5 tons, swung around and struck the side of the tank, ripping a gash in it that allowed the liquid hydrogen and oxygen to gush out and mix. The inevitable result was the fatal explosion.

The boosters had originally been designed as seamless, one-piece affairs, but, since they were just under 150 feet long and 12 feet in diameter, the manufacturer had cut them into sections for easy transportation to the Kennedy Space Center at Cape Canaveral. There they were reassembled, much like giant sections of pipe, and the joints were sealed with flexible synthetic rubber O-rings designed to expand under the pressure of the burning rocket fuel and keep the joints tight. The joints, officially called *field joints,* were triple-sealed to assure safety. First there was a layer of zinc chromate putty fortified with asbestos fibers between the mating pieces of the sections and the inner insulation that protected them from the burning rocket fuel. Then came the two O-rings, staggered one after the other. The joints were held together by 170 metal pins around the circumference of each. The system had worked well in all previous shuttle flights—there had been no accidents.

But this was not an ordinary flight. Problems and delays had plagued it from the start. Originally scheduled for July 1985, this flight, officially designated as STS-51L, was six months behind schedule. The White House was growing impatient. Congress, which authorized NASA's budget, was growing impatient. NASA's civilian customers—companies that wanted satellites placed into orbit or retrieved—were growing impatient and starting to look elsewhere for a carrier. NASA wanted very much to be mentioned in President Reagan's State of the Union message, which was scheduled to be delivered on January 28. It is understandable that NASA thought it urgent that this flight should not be delayed.

NASA had counted on a more generous time margin, scheduling the flight for January 22. It was pushed off one day, then two days, because of mission difficulties. On January 25, it was scrubbed again because of bad weather at the transatlantic abort landing site at Dakar, Senegal, on the western bulge of the African continent. A prediction of bad weather at Canaveral on January 26 caused another day's postponement. Then the next day's opportunity was lost when a hatch closing fixture for ground servicing equipment could not be removed from the orbiter. The offending fixture had to be sawed off and a bolt drilled out before the hatch could be closed. Meanwhile, crosswinds at the launch site made everyone nervous. On January 28, there was another two-hour delay when a component of the fire safety system failed and had to be fixed. Not until 11:38 A.M. did the *Challenger*'s mighty engines roar and lift it from the launching pad.

As mentioned, it was a chilly morning in South Florida. The ambient temperature stayed stubbornly at about 38 degrees Fahrenheit, and there was ice on the aft right field joint of the right rocket booster where the sun could not penetrate. There was ice, too, on the launching pad and on the bridge along which the astronauts walked from the service-tower elevator to the crew compartment. The NASA engineers protested. It was not safe to launch in such cold weather, they said. The cold would make the O-rings stiff and inflexible, and the joints would leak. The engineers from Thiokol, which had manufactured the boosters, agreed. They were overruled by NASA's brass and the Thiokol management, which was anxious to accommodate an important and impatient customer.

A special commission, the Rogers Commission, was created by executive order on February 3, 1986, to investigate the causes of the disaster. The famed physicist Richard Feynman testified before the commission. When it expressed skepticism about the O-ring failure, Feynman pulled out of his pocket an ordinary O-ring, the kind used in household plumbing. He asked the commissioners to squeeze it. The O-ring was springy and flexible. He then requested a glass of water with ice cubes and dropped the O-ring into it. After a few minutes, he fished out the O-ring and passed it around again. The commissioners found, to their astonishment, that it was now stiff as a board. Case proven.[27]

Feynman also submitted his personal report as an appendix to the commission's report. In it, he took NASA severely to task for

[27] It was not Feynman who came up with the idea of O-ring failure. In 1985, an engineer with Morton Thiokol, Roger Boisjoly, noticed a large amount of blackened grease between the two O-rings during the postlaunch inspection of recovered boosters from the third shuttle flight, 51C. However, he was unable to convince NASA that there was a serious problem.

arrogantly assuming that, because every launch had gone well in the past, all future launches would proceed without mishap. The problem of O-ring erosion had occurred before, he pointed out, and NASA had done nothing about it. It was cheaper not to replace the huge 12-foot seals. And on previous flights the O-rings had still functioned. But this time they did not. Feynman recommended that future shuttle designs be conducted "from the bottom up" by the engineers, who would test materials and components thoroughly before putting them together. NASA seemed to prefer doing things "from the top down"; that is, management decided on the final product and left it to the engineers to find solutions. This is the method often used in the auto industry, which is dogged by well-known inefficient and trouble-prone designs.

As a result of the *Challenger* disaster and the ensuing investigation, the space shuttle program was abandoned for several years. It did not resume until 1991, with, presumably, much stricter precautions.[28]

* * *

Six hundred and seventy-six human lives were lost in the crashes of TWA 800, Swissair 111, and EgyptAir 99, and seven were lost in the *Challenger* disaster. Could any of these tragedies have been averted? The painstaking detective work of the transportation safety boards of two nations—the U.S. and Canada—indicates that stricter inspection might have prevented the disastrous fire on board TWA 800, while Swissair 111 might well have been landed safely if the senior pilot had not insisted on going by the book. (However, in late March

[28] As a tragic footnote came the *Columbia* disaster of January 16, 2003. Again engineers had issued warnings to management and again had been overruled. In this case, the problem was not leaky O-rings (for even the top officialdom was by now aware of their dangers and not anxious to repeat the *Challenger* experience) but apparently one or more loose pieces of the insulation that was supposed to protect the shuttle during its reentry into the atmosphere. Again ice may have been the culprit: moisture soaking into the foam of the insulating tile may have frozen, expanded, and pried the tile loose from the shuttle's wing. Whatever the cause of its loosening, the tile, decelerating while the shuttle was rapidly accelerating, would have acted like a projectile. But since the *Columbia*, or more correctly its pieces, came down over land, the investigation of the disaster does not fall under the heading of underwater detective work.

2003, the final report of Canada's Transportation Safety Board concluded that, once the pilots became aware of the on-board fire, it had already progressed to the point where the Swiss jetliner could not have been saved.) As for EgyptAir 99, we will never know exactly what happened, but the flight would almost certainly have been completed normally if Gameel al-Batouti had not been aboard. The midair destruction of the space shuttle *Challenger* would probably not have occurred if NASA, in servile eagerness to present President Reagan with a sound bite for his inaugural speech, had not placed political considerations above safety. If the NASA brass had only listened to their own engineers and waited for the air at the launching site to warm up, the worn-out O-rings might have held together for one more flight. And if NASA had been willing to dip into its budget to replace the O-rings, that particular failure would not have happened. If ... if ... and if.

9 Peril in the Silent Service: The *Kursk* and Other Submarine Tragedies

There are probably worse ways to die than aboard a damaged submarine, but it is still a horrifying prospect: the slow suffocation as your oxygen runs out, the painful headaches as carbon dioxide builds up, the chill of the water as it creeps ever higher inside the wounded hull, and, above all, the knowledge that you are doomed. Of course, your death may be swift rather than slow—an explosion may snuff out your life in an instant, or water pressure may crunch the hull like an empty soft-drink can if the sub dips below its crush depth. Or you may die in the fires that spring from electrical short circuits (a problem to which Russian nuclear submarines are especially prone) or be asphyxiated by toxic fumes. The luckless crew of the *Kursk*, pride of the Russian Navy, experienced at least two of these possibilities.

On August 10, 2000, the *Kursk* set off on a training exercise with Russia's Northern Fleet. Aboard were 118 men, including five high-ranking naval observers and two Russian sub designers. Two days later, the submarine was lying on the floor of the Barents Sea, some 354 feet down, with all her crew and passengers. None survived. The disaster made headlines and TV specials around the world and caused a major surge of outrage in Russia.

The *Kursk*, launched in 1995, was a giant among submarines. She measured 508.53 feet from stem to stern—nearly twice the length of a football field. Her hull enclosed ten compartments. Had she been stood on her bow at the spot where she sank, her stern would have towered fifteen stories above the waves. Her beam was just under 60 feet, and she displaced 26,400 short tons of seawater when fully submerged. Her two nuclear reactors, which fueled twin steam turbines of 49,000 horsepower each, drove her at a top speed of over 30 knots (the exact figure is classified). Following Soviet and then Russian practice, she was double-hulled, with a relatively thin outer hull shielding the heavy inner hull. A 4-inch-thick rubber layer coated the outer hull and was designed to render the submarine less visible to sonar.

Several of the ten compartments inside the hull were used for storing munitions. Designed as an attack submarine with the mission of destroying enemy (particularly American) aircraft carriers or bombarding land-based targets, the *Kursk* carried twenty-eight torpedoes and twenty-four cruise missiles, each capable of carrying a nuclear or a conventional warhead.

Another pride of the Russian submarine fleet was the *Shkval* (*Squall*) torpedo. Rocket-propelled, it could streak through the water at 230 miles an hour thanks to a semi-vacuum bubble that it created around itself. The range was reportedly 43.5 miles. It is believed that some of the *Kursk* armament were *Shkvals*. This state-of-the-art submarine also carried two practice torpedoes with dummy heads, designed to hit a target and then be recovered. Starved for funds, the Russian Navy outsourced the manufacture of these torpedoes to a factory in Bishkek, the capital of Kyrgyzstan in Central Asia. The practice torpedoes were substantial, measuring 29.5 feet long and weighing about 1,320 pounds apiece. They were not something that could be handled without mechanical hoists.

The *Kursk* was firing practice torpedoes on the second day of the exercise when, suddenly, around 11:30 A.M., there were two powerful explosions two minutes and fifteen seconds apart. They were picked up by the nearby Russian flagship *Pyotr Velikiy* (*Peter the Great*), by another Russian sub maneuvering in the area, by an American spy ship trailing the Russian fleet at a discreet distance, and by a number of seismic listening stations on land. The second explosion, by far the more powerful, was detected as far as 3,100 miles away. But nobody knew what the blasts were.

The first clue that something had gone wrong aboard the *Kursk* was her failure to report that her mission was completed and that she had vacated the target area. The second was when she did not respond to an order sent out six hours later to report her location and condition. At 11:30 P.M., after the prescribed wait of twelve hours, the Russian Navy announced that the missing sub had suffered an emergency. At 4:36 A.M. the next morning, the sonar of the cruiser *Pyotr Velikiy* located the *Kursk*, which was lying immobilized on the sea floor 354 feet down. The "sighting" was confirmed that evening by a submersible from a Russian rescue ship.

But rescue efforts did not begin immediately. The first reaction of Russian officialdom was to institute a cover-up. Even as President Vladimir Putin declined to interrupt his vacation to handle the crisis, the Russian government was pumping out optimistic press releases about rescuing the crew. Behind the scenes, high-ranking naval officers began an orgy of accusing each other of incompetence and negligence. Meanwhile, the ruined *Kursk* sat on the sea floor, her crew helplessly trapped.

At first, the Russian establishment resisted the idea of enlisting foreign aid to rescue the crew of the *Kursk*. To call in foreigners, they feared, might imply that Russians were not competent to do the job themselves. In the end, however, they had to call in British and Norwegian rescue teams (although the United States offered help, it was turned down, presumably for fear of spying).

In the meantime, the Russians mounted a strong propaganda attack against the United States, claiming that a U.S. submarine had rammed the *Kursk*, fatally damaging it. This accusation was actually not at all far-fetched: U.S. and Soviet/Russian nuclear subs had been spying on each other for years, often at very close quarters. And it happened now and then that, if one of the subs made an unexpected turn, the other one would be unable to avoid colliding with it. Moreover, the Barents Sea, with its proximity to the bases of Russia's proud Northern Fleet, was a major area for fleet maneuvers—and for spying.

(One such incident occurred in the North Pacific in 1970, when the U.S. nuclear-powered attack submarine *Tautog* collided with a Soviet nuclear-powered sub that it was shadowing, the K-108. Apparently, the Soviet captain, suspecting he was being tailed, made a high-speed turn and circled back. His sub passed right over the *Tautog*, scraping against its sail. According to one American web site devoted to submariners' reminiscences, the K-108 bent its propeller shaft in the collision, ruining the seal where the prop shaft left the hull. Seawater under tremendous pressure poured in uncontrollably, and the K-108 imploded. The American crew listened in horror as they heard the noises of the other craft breaking up and sinking, feeling themselves lucky to be alive. A less dramatic version of the incident states that the K-108 survived the collision but suffered a hole in her outer hull and left part of her starboard propeller in the *Tautog*'s sail. [The starboard propeller shaft was indeed bent.] Somehow the K-108 made it back to port. The *Tautog*, believing that the Soviet sub had sunk, returned to port at Pearl Harbor for repairs to its sail, periscopes, and antennae.)

Other theories were avidly propounded. The *Kursk* had hit one of the wrecks that littered the sea floor (shattered ships from World War II and later, and unlucky trawlers, abounded). Or it had blundered upon a leftover mine and blown up. Or it had been fired upon by the heavy cruiser *Pyotr Velikiy*, the command ship of this training exercise.

For public consumption, Russian officialdom issued optimistic reports on the safety of the trapped crew and passengers of the *Kursk*. The design of the sub would protect them, said the government functionaries, and soon they would be rescued with the aid of a diving bell–like device that could mate with the sub's escape hatch. Perhaps the *Kursk* was not even so severely damaged that her skilled and dedicated officers and crew could not extricate her from her sea-floor nest in the clinging mud. But one day passed, and then two. It was obvious that oxygen inside the sub was running out. Faint tappings were heard from the rear section of the ruined sub, giving hope that at least some of the crew were clinging to life. The sounds continued for three or four days, but later the authorities admitted that they were "mechanical," caused by the machinery disintegrating further.

The Russian Navy did not sit idle while the public relations people were creating smoke screens. When the *Kursk*'s location was established, attempts were made to connect an air hose and an electric power line from the surface. They failed. Then the deep submergence rescue vehicles (DSRVs) were called into action. At least eighteen dives were made to the sunken submarine, but their entry portals could not mate with the rear escape hatch of the *Kursk*—it was too badly bent out of shape, either by the explosion or by the impact when the sub's hull struck the sea floor.

A Norwegian rescue team with its own rescue vehicle was called into action. It, too, found it impossible to make a proper docking with the *Kursk*. Meanwhile, Russian and Norwegian divers, getting no response from their attempts to communicate with the trapped crew by banging on the hull, concluded that no one was left alive. Although details are not available, we can assume that these divers were using heated suits and breathing a special mixture of gases adapted to the 354-foot depth and the extreme cold.

A British rescue sub, the LR-5, was rushed to the scene. The 30-foot craft could take sixteen rescuees at a time, though each trip would require two and a half hours in the case of the *Kursk*. The LR-5 was equipped with a useful innovation: a pivoted hatch in its bottom that could be angled to match the tilt of a sunken submarine (the *Kursk* was lying at an angle of about 20 degrees to starboard). A flexible skirt was supposed to give a watertight seal around the wrecked sub's escape hatch. But the British sub arrived too late, just as the Russian Navy announced that there was no hope of finding survivors.

The brief Arctic summer with its relatively calm weather was drawing to a close. Any efforts at salvage would have to cope with strong winds, high seas, and storms as the season changed. It was manifestly impossible to attempt to bring up the sunken *Kursk* with its nuclear reactors and load of unexploded munitions before the next summer at the earliest, and a contractor had to be selected and payment agreed upon. And these matters take a long time and much bargaining.

In the meantime, the depressing work of recovering the bodies of the dead crew went on. Divers, using high-pressure water jets loaded with diamond dust, cut a big hole in the sub's triple-layered hull to gain entry to the eighth and ninth compartments, those farthest away from the blast. One diver's search was cut short at 20 feet into the hull because his air hose would go no farther. But the first four bodies were pulled out on October 25, to the accompaniment of national mourning. Four more bodies were recovered before bad weather put a stop to operations for the long Arctic winter.

One of the bodies belonged to Lieutenant-Captain Dmitri Kolesnikov. In one of its pockets was a note scrawled on a page torn from a detective novel. Public opinion demanded that the note's contents be made known, so the deputy prime minister read selected bits from it at the funeral service for Kolesnikov and his three comrades.

"13:15 [1:15 P.M.].... All the crew from the sixth, seventh, and eighth compartments went over to the ninth. There are 23 people here. We made this decision as a result of the accident. None of us can get to the surface.... I am writing blindly.... There is no need to despair. It is dark to write but I will try by feel. It seems there is no chance, let's hope for 10 to 20 percent, that someone will read this.... Here are the lists of the personnel of the sections who are in the ninth and will try to get out. Hello to everyone. There is no need to despair," said the gallant note. Apparently, the rest of the note was deemed too politically embarrassing to be made public.

Kolesnikov's handwriting began neatly and legibly, as if he had plenty of light, but evidently the batteries in his flashlight ran down, and he had to scrawl the last bits "by feel" in the dark. Kolesnikov's note gave the lie to an earlier official statement that all the crew had died within minutes of the explosions.

Later it was announced that another crewman's body also bore a note, which read, in part, "... There are 23 people in the ninth compartment. We feel bad.... We're weakened by the effects of carbon monoxide from the fire ... the pressure is increasing in the compartment ... if we head for the surface we won't survive the compression."

After the postmortem examination, investigators announced that the cause of death was primarily carbon monoxide poisoning. The faces of the dead men were flushed—a telltale symptom of the condition. And when the examiner pressed on their breastbones, they made a crackling sound, and "a foamy liquid dribbled out of their nostrils." These last two phenomena indicated that the men had been breathing high-pressure air before they died, and their blood had become saturated with nitrogen. Conclusion: the air in the compartment in which they were trapped had been compressed by the incoming seawater.

The bodies of the rest of the crew had to remain in a sort of cold storage in the ruined hull until salvage operations could begin the following year.

After a great deal of haggling and maneuvering, accompanied by clouds of self-exculpatory disinformation from various government agencies and officials, a deal was struck with two experienced Dutch salvage firms, Smit-Tak and Mammoet. Smit-Tak would do all the underwater work, and Mammoet (which means *mammoth*) would do the actual raising.

The salvage operation was a marvel of technology and preparation, with every step worked out months in advance. It is worth following in detail.

The plan was to cut off the damaged bow section of the *Kursk* with a sort of gigantic underwater chain saw. Then grippers would be installed in twenty-six holes that divers would cut in the top of the hull, and huge hydraulic jacks mounted on a semi-submersible barge, or pontoon, would lift the *Kursk*, ever so slowly and gently, from its bed in the sea-floor muck. Finally, pontoon and submarine would be towed to a floating dry dock at the great Russian naval base in Murmansk for the forensic examination.

As early as June 28, 2001, the specially designed chain saw was tested in Rotterdam. It was made up of a series of abrasive cylinders connected by a chain and operated by a power source at each end. The designers tried it out first on a retired dredge being cut up for scrap and then on a piece of steel of the same type used in the *Kursk*. It passed the test with flying colors.

The lifting pontoon, *Giant 4*, had to be modified so that the conning tower and tail fin of the *Kursk* would fit snugly underneath, and the jacks had to be installed. This was a seven-week process, so the *Giant* was not due at the salvage site until late September.

The first phase of the salvage operation was to survey the site where the *Kursk* was lying and to blast away some of the muck surrounding the hull with high-pressure hoses. This process was daintily described as "soil washing." Then Smit-Tak's divers began to attack the hull with high-pressure abrasive blasters. The sub's designer, Igor Spassky, a professor from the Rubin Central Design Bureau, had instructed them precisely where to cut the holes so as not to weaken the hull or disturb compartment No. 6, where the nuclear reactors that had once powered the ship were located. Because of the pressure at the depth where the *Kursk* lay, the divers had to spend weeks in a pressure chamber when they were not at work on the sea floor. A sort of diving-bell-elevator shuttled them between the work site and the pressure chamber on the deck of a support ship.

Because the steel of which the *Kursk* was made was so hard, the work was slow. The initial holes were only 4 inches in diameter; the divers widened them in increments of fractions of an inch at a time until they were large enough to accommodate the grippers (usually described as *plugs* in news bulletins). The grippers were large steel devices whose business ends boasted a pivoted, flanged arm on either side; their opposite ends were attached to lifting cables. A hydraulic cylinder on each gripper would force the arms outward once the device was well inside the inner hull of the sub, and the protruding flanges would give it purchase.

While this was going on, a ship called the *AMT Carrier* arrived on the scene with two suction anchors. These were designed to burrow into the sediments, one on either side of the sub, and guide the chain saw as it worked its way from the top of the hull to the bottom. The ship also carried the saw. Although the *AMT Carrier* was held up by bad weather, on September 4 the sawing had begun. The short window of opportunity meant round-the-clock operations, so the first cut was scheduled for midnight.

The saw worked admirably for two and a quarter hours, slicing a quarter of the way down through the *Kursk* before the equipment failure that plagues undersea operations struck. The cables, worn by stones and abrasive sediment, broke and had to be replaced. This happened several times before the job was done.

The *Giant*, towed from Rotterdam all the way around the North Cape, arrived on site a week late because of bad weather. Once in position, it was held in place by eight huge anchors. Stormy weather held things up again for a couple of days. As soon as sea conditions permitted, divers guided the grippers into their holes; the expansion arms were forced into place, and the delicate operation of lifting began.

Each of the *Giant*'s jacks operated a bundle of fifty-four high-strength wire cables. Devices on each jack compensated for the heaving of the vessel on the seas and kept the force on the attachment points constant. Everything was controlled by computers with massive backup systems. It was vital to keep from stressing the *Kursk*'s hull, since just where and how much it had been weakened by the second explosion were unknown.

The salvagers faced another problem: the sticky nature of the sediment, which one Mammoet official likened to chewing gum. The experts feared that the suction of the sediment on the *Kursk*'s hull would hold it down until it came free with a terrible jerk only to plummet back to the bottom and play havoc with the lifting machinery. And what would happen if the nuclear reactors sprang a leak or came loose?

A simple technique solved the problem. By alternately lifting and slacking off on the cables on opposite sides of the sub, the crew were able to rock the hull back and forth until, after three hours, it came free. Inch by inch, the ponderous hull was raised until it nestled firmly under the *Giant*. Then the gaping hole where the bow had been cut off had to be sealed with a patch. At last, two powerful tugboats towed the *Giant* and its burden to a floating dry dock in Roslyakovo, near Murmansk.

The next stage of the operation began with Russian workers clearing out the interior of the ruined submarine after the corpses of the dead crew were removed. They found a wilderness of twisted and torn metal and melted plastic and removed over 440 short tons of exploded ammunition. The workers also recovered five automatic recorders, like the black boxes of airplanes, and the sub's log books; none of their contents has been disclosed. (In fairness to the Russians, this secrecy is a universal practice among military establishments—they like to keep their mistakes confidential.)

Then an enterprising Russian journalist turned up an unpleasantly surprising fact: the body of the gallant Lieutenant-Captain Kolesnikov had a broken nose and broken ribs. Apparently, the trapped sailors had lost all restraint and attacked the only officer present, Kolesnikov. Officers who had personally read his note said that he had described the mutiny. The Russian Navy, following the self-preserving instincts typical of bureaucrats of all nations, had suppressed this negative news.

What did cause the explosion that sank the *Kursk*? A forensic reconstruction of the accident concluded that a practice torpedo had exploded in its firing tube, starting a flash fire that blew up most of the munitions stored in the bow compartments. To begin with, the Norwegian divers who had tried to rescue the crew found a huge hole in the bow, over 21 feet long, and the metal was bent outwards, indicating that the blast had come from inside the sub. Water instantly rushed into the holed first compartment at an estimated two to three tons *per second*. In the short interval before the second explosion, the *Kursk* had taken on some 300 short tons of water in her bow—while not necessarily a fatal amount of water, it was enough to send her to the bottom.

Instances in which torpedoes had exploded prematurely in other submarines were reviewed. Several naval officers told the press that the hydrogen peroxide–kerosene mix that was used to propel the practice torpedoes would react explosively upon contact with copper or brass—and the *Kursk*'s two practice torpedoes were made of brass. The procedure was to fill each practice torpedo with the fuel mix just before firing it. A slip of the hand could mean disaster.

Naval records showed that the torpedo room was under the charge of one midshipman and two naval draftees. One of the conscripts had written home shortly before the *Kursk* left port on its final voyage to say that he expected a promotion if he performed well firing the torpedoes; this was a strong indication that he was still in his first year of service and inexperienced. The midshipman also had to be young and unseasoned, and perhaps the other draftee was, too. Many believed that the enthusiastic young conscript made a mistake in loading his torpedo, with fatal results.

Stories also appeared in the Russian press that one of the practice torpedoes had been dropped while being loaded onto the train for its long journey from Kyrgyzstan to Murmansk and suffered some distortion. The damage could make it stick in its tube when fired. The crew, said the stories, found out about the damaged torpedo and complained—to no avail. For what it is worth, the Russian Navy abandoned this model of practice torpedo after the disaster. Although Russian naval officers generally agreed that the tragedy was instigated by a misfired torpedo as early as a week after the accident, it was not until the summer of 2002 that it was officially acknowledged as the cause.

Despite the inrushing water, the flames from the torpedo could not be contained, and two minutes and fifteen seconds later the rest of the munitions blew up, instantly killing all but one of the officers and most of the crew. The twenty-four survivors of the blast perished soon afterward, unable to get out through the escape hatch.

In a final note of disgust, the Russian journalist who had uncovered the details of the mutiny on board the *Kursk* quoted her source as saying that, in his opinion, the commander of the submarine knew that he had a defective torpedo on his hands—but was forced by his superiors to fire it anyway, for he was not to deviate from set plans.

* * *

It was nearing the close of January 1968 when the Israeli submarine *Dakar* vanished mysteriously in the Mediterranean. The *Dakar* was a rebuilt British submarine of World War II vintage, sold to the Israeli Navy. She was en route to her new home from the shipyard in Scotland where the renovation work had been done when she disappeared.

The *Dakar* was built in 1943 and named HMS *Totem*. In the mid-1950s, she was rebuilt extensively: she was lengthened by 12 feet and given a new, streamlined conning tower and bridge, and the deck gun was removed.[29] Then, in 1965, the *Totem* was sold to Israel and

[29] Up to the end of World War II, submarines were designed primarily as surface ships that could submerge since the electric batteries on which they relied for underwater operations had a limited capacity. The deck gun was a useful weapon for threatening—and sinking—unarmed civilian ships. But after World War II, with the German invention of the snorkel coming into wide use, subs could cruise underwater at shallow depths on diesel power for many hours, and they were designed to fight underwater. The deck gun could not be fired underwater and became only a source of drag.

underwent a two-year reconstruction program in Scotland. There she was fitted out with a modern navigational system and given a huge conning tower-and-bridge (often called a *sail*). The conning tower, which measured over 19 × 15 × 9 feet, was designed as an exit/reentry chamber for a team of ten commando raiders whose targets were terrorist bases along the Mediterranean coastline. The conning tower was made of aluminum, which is sometimes used in ships and ROVs because of its light weight. Aluminum can cause problems, however. Together with steel in the presence of seawater, it sets up an electrochemical reaction that corrodes it away. This corrosion must be prevented by keeping the aluminum component meticulously covered with paint so that it does not come in contact with the highly conductive seawater.

When the work was done, the sub measured 285 feet in length. Twin diesels gave her a surface speed of 16 knots, and two electric motors pushed her along at a top speed of 10 knots when submerged, a remarkable pace when one considers the drag that her towering sail must have caused. She carried six torpedo tubes, air- and sea-scanning radar, and a towed underwater radar system.

Upon completion of the work, the submarine was turned over to the Israelis, who renamed her *Dakar*, the Hebrew word for *swordfish*. The *Dakar* was one of a class of three converted British subs; her sisters were named *Dolphin* and *Leviathan*. After sea trials, the *Dakar* left from Portsmouth, England, headed on the long journey to her new home port of Haifa in Israel. Protocol called for her to radio her position to naval headquarters once a day and send a coded control message, which did not reveal her position, every six hours after that. When she reached Gibraltar, she came into port for most of the day and then left at midnight. This was the last time that the *Dakar* and her crew were seen.

When the *Dakar* had passed Crete, the radio message included a serenade by the crew, who sang a lighthearted song about their boat composed by one of the officers. There were three more coded messages ... then nothing. The messages ceased at 12:02 A.M. on January 25, 1968.

An international search-and-rescue (SAR) mission was launched the following day; the Israelis received help from the U.S., the U.K.,

Greece, Turkey, and even Lebanon. Since a radio station on the island of Cyprus had picked up an SOS call on the missing sub's frequency, the searchers concluded, based on her planned route, that the *Dakar* had sunk somewhere southeast of Cyprus. After six days, the fruitless search was called off by the international helpers, though the Israeli Navy continued it for four more days.

Ironically, the *Dakar* had made such good speed that she neared Israel well ahead of her scheduled arrival date of February 2. Commander Ya'acov Ra'anan had radioed headquarters for permission to dock early. HQ told him to come into port on January 29. Ra'anan then asked if he could come in on January 28. The answer was no, because a big welcome ceremony had already been planned. That ceremony never took place.

Proper burial of the dead is very important in Jewish religion and tradition, and Israel spares no effort to recover the bodies of servicemen and -women who have died in action. The Israeli government sent out twenty-five search expeditions over the next thirty years, covering virtually every part of the eastern Mediterranean. The only area that they omitted was along the planned route of the *Dakar*.

But they had been misled by the discovery of an emergency buoy from the *Dakar* that had washed up on the beach near Khan Yunis, an Arab fishing village at the southwest end of the Gaza Strip, a year later. The *Dakar* carried a pair of these buoys, attached to her hull by 656-foot-long metal cables. In case of an emergency, the crew could release the buoys, which would float to the surface and send out an SOS message for the next forty-eight hours. The buoy that washed up still carried a short length—about 2 feet—of its cable. Israeli scientists examined the buoy and cable exhaustively and concluded, on the basis of the scant evidence they had, that the buoy had remained attached to the missing sub until it broke off. The *Dakar*, they decided, was resting in a spot between 492 and 1,069 feet down and 50 to 70 nautical miles off her planned course.

Eventually, other scientists took a look at the buoy and the algae that grew on it and discovered that they did not correspond to the algae native to the Khan Yunis area. The search was on again, this time aided by the U.S. Navy.

The Israelis hired a Maryland-based American firm, Meridian (now called Nauticos Corporation), to locate the wreck, and the quest began in the spring of 1999. Personnel from the Virginia-based firm Phoenix Marine Incorporated and the Seattle-based firm Williamson & Associates were also on the team.

Meridian selected the one area that had not been previously investigated: a section of the *Dakar*'s scheduled course. This area would have to be east of the *Dakar*'s last radio transmission—but how far east? Factoring in the assumed speed of the sub with classified U.S. Navy data on Mediterranean currents, the team staked out a 60 × 8 nautical mile rectangle more or less west-southwest of the island of Cyprus. Two support ships "mowed the lawn"; one carried the side-scan sonar and the other an ROV. With the search area divided into sixteen search lanes, it took the ships, motoring at 2 knots, between thirty and forty hours to complete each pass, with the sonar fish hovering between 250 and 325 feet above the sea floor. On the evening of May 24, after almost a month of searching, the sonar picked up a large body on the sea floor. The sonar image showed a large black spot with several smaller dots scattered around it. Tom Dettweiler, the leader of the expedition, ordered the ROV to be lowered—but at that point bad weather set in and held up the operation for three days. Not until 7 A.M. on May 28 could the ROV be sent down.

The wait paid off—the ROV sent back clear pictures of the wrecked hull, with the conning tower ripped off and lying on its side. The bridge, once part of the conning tower, lay more than 750 feet away. The periscope was in its up position, indicating that the sub had been cruising near the surface when disaster struck. The hull itself was broken in two: the stern had broken off just aft of the engine room and was lying at an angle to the rest of the hull. The engine room was a wreck, but the emergency buoy at the bow was still snug in its wooden box.

The long-missing *Dakar* was lying at a depth of approximately 9,500 feet and only 3 miles off its scheduled course—excellent positioning for the navigational equipment used in 1968. So what caused the submarine to sink?

Over the years, several theories had circulated. One was that Commander Ra'anan, with two extra days on his hands, had decided to take his ship into Egyptian waters to do a little unauthorized reconnaissance. While there, he had been spotted by the enemy and sunk. Indeed, over the years several Egyptian naval officers had mendaciously claimed to have personally sunk the *Dakar* variously by gunfire, torpedoes, or ramming. (In 1998, an Egyptian admiral officially announced that Egypt had destroyed the Israeli submarine.) Or perhaps the Soviets, who at that point were mentoring Arab military forces, had sunk the *Dakar* themselves. Or Ra'anan, knowing that he had been spotted by the Egyptians, ordered a crash dive that went wrong. These theories had some plausibility, for Egypt was seething with hatred of and resentment toward Israel after its ignominious defeat in the Six-Day War of 1967 and could have been expected to take vengeance on an Israeli ship in Egyptian waters. On the other hand, the listening devices with which various governments embroidered the sea floor had recorded no explosion.

Yet another theory held that the British had sold the Israelis a mechanically defective craft. In fact, Ra'anan's messages had complained of mechanical problems and minor leaks. But perhaps the most persuasive of all the theories was that the *Dakar*, cruising at periscope/snorkel depth, had been run down by a very large tanker or bulk carrier, which would not even have noticed the collision. Even if the submariner on watch had seen the big ship—and submarines have blind spots—the sub might not have had time to dive to a safe depth.

After the discovery of the *Dakar* in 1999, the experts interpreted the photographic evidence, particularly a large gash in the stern, as pointing toward a collision as the cause of sinking. But the evidence of the next year's expedition, with much clearer images, showed no obvious signs of such an impact.

One of the first things an investigative team does is to look for signs of a fire on board the vessel. Through the video "eyes" of the ROV, the Nauticos team checked the engine room, control space, and crew quarters for signs of blackening and blistered paint. They found none. They did find that the valves of the great storage batteries that powered the sub when fully submerged were open. The valves would have been opened to vent gases produced while the batteries were being charged, and so the investigators deduced that before it sank the sub was running on its diesel engines just beneath the surface, with its snorkel deployed.

The external pipes, such as the ballast pipes and cooling-water pipes for the engines, were intact, not crushed. This lack of damage could only have resulted if the water pressure inside and outside the pipes remained equal as the *Dakar* sank, and this in turn meant that their valves had been open—another indication that the submarine had been cruising at shallow depth. Furthermore, the watertight bulkheads between the ship's compartments were not closed off, as they would have been in preparation for a dive. The speaking tubes were open. Nothing appeared to have been done to prepare the *Dakar* for a dive.

The dean of marine forensics, Robin Williams, was with the Nauticos group. From the thickness of the hull's metal and the way in which it was bent, Williams deduced that the sub had gone from cruising depth to its crush depth of some 800 feet in less than thirty seconds, giving the crew no time to take action. In training, under ideal conditions and with the crew knowing what is expected of them, it takes two and a half to three minutes for them to get into action. In real life, of course, an emergency takes everyone by surprise. As Williams said, the crew were probably doing their best to save the sub when, suddenly, "it was as if someone turned off a light."

The investigators found two different types of failure in the main hull, which occurred almost simultaneously—a very rare event. One was the expected collapse from the external pressure of the sea, and this was in evidence from the engine room as far forward as the crew's quarters. But in the control space, something very different took place. Under the floor were very heavy saddle tanks, and they exerted such a load on the circular frames of the hull as the sub went out of control that they gave way, pulling the hull plates inward as they buckled and nearly turning the ship inside out.

The very forward section of the submarine was not collapsed. This proved that it had been completely flooded before the ship went down. (The pressure of the water inside the sub resisted the pressure of the water outside it.) One cause of the flooding may have been a leaky torpedo tube, the leak exacerbated by the relatively high speed at which the *Dakar* was cruising, which was 8 to 8.5 knots.[30] Or the water might have leaked down through the snorkel tube. Snorkeling is always hazardous for a submarine; a wave can easily overtop the snorkel tube and penetrate to the interior. Bit by bit, the water can accumulate to the danger point without the busy crew noticing it. Whatever the cause, the ship suddenly became nose-heavy and drove itself forcefully down to its crush depth.

The bridge, or forward section of the sail, as noted above, lay some 750 feet away from the rest of the wreckage. The experts found that it had been blown clear because its pressure tank had collapsed at a much greater depth than the rest of the vessel. This implosion, paradoxically, had triggered an explosion of the air trapped inside and blown the forward part of the sail clear as the *Dakar* sank. The explosion was so powerful that some of the sail's tough steel was bent into curlicues.

The sail, weighing 3 or 4 tons, was salvaged and returned to Israel to become a monument to the lost sub and her crew of sixty-nine, along with a gyrocompass from the bridge and a blender from the galley. The Egyptians' boastful claims were proven false, and Commander Ra'anan could no longer be blamed for deviating from his orders.

The recovery of the sail was considered a technological feat in its own right. Nauticos gave much credit to the state-of-the-art rope it used, called *Dyneema*, made from an extremely high-strength, positively buoyant polyester fiber. The expedition's ROV was sent down with the rope coiled in figure eights in its recovery basket. Once the ROV had reached the sail, it set the basket down, and the pilot maneuvered it over to the sail to attach T-clamps, designed so that, the harder one pulled on them, the more tightly they gripped. The ROV then attached the rope to the clamps, a delicate operation

[30] This speed was calculated from the distances between the daily location points that the *Dakar* radioed in to headquarters. Tom Dettweiler speculated that Captain Ra'anan had been trying to set a speed record for submerged travel with his new submarine—submariners are a competitive lot, he said.

(imagine tying knots at a distance of nearly 2 miles without even being able to feel what you are doing) that took several hours. Then the winch was eased into motion for the lift; the clamps held, and the sail began a nine-hour journey back to the surface, where it was lashed securely to the deck of the support ship for delivery back to Haifa. The recovered sail is now on display in a special museum.

* * *

Although submarine design has advanced by quantum leaps since the days of World War I, accidents still occur with distressing frequency, even on nuclear-powered submarines. Russia has the worst accident record, followed by Britain and France. Most of these accidents were caused by fires. The United States has the best record, but even so it has lost two nuclear subs at sea.

The first was the USS *Thresher*, which sank during sea trials off the coast of New England on the morning of April 10, 1963. The 278.5-foot submarine was almost new. Built at the famous Portsmouth Naval Shipyard in Kittery, Maine, she was launched in 1960. Her mission was fast attack, and she was capable of 20 knots submerged. She was the first of a new class of submarines with significantly greater depth capability than any other subs in the U.S. Navy, and the accident happened when she was operating at her rated limit.

The *Thresher*'s early cruises, on which she tested a submarine rocket system, had been great successes. But in the fall of 1962, while mooring at Port Canaveral, Florida, she was struck by a tugboat, and one of her ballast tanks was damaged. After being repaired at the great submarine base in Groton, Connecticut, she returned to the Florida Keys for more tests. She spent the balance of the winter and early spring in dockyards.

On April 10, 1963, she set out from Portsmouth for deep-diving tests, carrying sixteen officers and ninety-six crew, plus seventeen civilian observers working for the companies that manufactured her specialized equipment. For the sake of safety, a submarine rescue ship, the *Skylark*, accompanied her. The *Thresher* duly descended to her rated crush depth, reporting this to the *Skylark* by acoustic telephone. But, according to one source, at 9:13 A.M. the crew of the

Skylark got an alarming message: "... minor difficulties, have positive up-angle, attempting to blow [the ballast tanks]." Then came a series of fragmentary, unintelligible messages. As they listened, horrified, they heard a noise like compressed air rushing into an empty tank ... then nothing more. That was at 9:18 A.M.

The rescue equipment that the *Skylark* carried, designed for 850 feet, could not reach the *Thresher*, which lay more than 8,000 feet down. She was declared missing that same day, with all 129 people aboard.

The bathyscaphe *Trieste* was pressed into service by the navy. It located the remains of the *Thresher* lying at 8,400 feet below the surface. The sub had been fragmented into six major pieces, surrounded by a field of smaller debris about 400 yards square. During a series of ten dives by *Trieste*, many photographs were taken to ascertain the cause of the sinking. Other photos were taken by a camera lowered from a surface vessel. A naval court of inquiry eventually concluded that a pipe joint in the seawater system that cooled the nuclear reactors had failed. Water under high pressure had spurted out, short-circuiting the electrical system. Without power, the ship was unable to blow its ballast tanks clear, the standard emergency response. The helpless *Thresher* rapidly sank below its crush depth and imploded. The crew and passengers were probably killed instantly by the inrushing water, which would have struck them like a gigantic blow.

Analysis of the photos and other evidence (probably fragments of the wreckage, though the navy doesn't say just what fragments nor how they were recovered) points to the probability that a number of the pipe joints were not properly connected. Specifications called for them to be silver-brazed by electrical induction heating, but the joints that were hard to reach were brazed with a handheld torch. This might not have given as strong a bond as was needed. The navy had experienced a series of near misses with failed silver-brazed joints before the *Thresher*'s accident and ordered that all joints be tested by ultrasonic sound. But ultrasonic testing proved cumbersome and time-consuming, and the shipyard had a schedule to meet, so it resorted to the older method of hydrostatic testing, which entails filling the pipes with water at increasingly higher pressures. All the joints passed this test, but apparently it was not enough.

Very much alert to the danger of radioactive material leaking from the reactors, the navy conducted a series of tests in the area, sampling the seawater, bottom sediments, and fish and other marine organisms. The tests showed that the level of radiation was not dangerous, thanks to the very safe construction of the reactors.

Learning from this tragedy, the navy adopted new inspection requirements for all submarines. The creation of DSRVs also stems from the loss of the *Thresher*.

The second American nuclear submarine to be lost was the USS *Scorpion*, in 1968. Submarine technology had evolved since the *Thresher* went down, and so had the game of undersea espionage that the U.S. and the Soviet Union were playing against each other. The *Scorpion*, unlike the *Thresher*, was operating under strict silence. She is believed to have been engaged in a top-secret navy plan to listen in on a Russian nuclear submarine exercise near the Canary Islands. Having collected the desired information, she was then to race home to Norfolk, Virginia.

The 252-foot *Scorpion* was one of the *Skipjack* class, designed specifically for anti-submarine operations. The modified teardrop shape of her hull reduced drag and gave her extra speed and silence. Her forward diving planes, mounted on her sail instead of at the bow, reduced interference with the bow sonar from water turbulence. Her crew of ninety-nine included a team of Russian-language experts to translate the conversations that the *Scorpion*'s advanced listening system picked up from the craft she was spying on.

The *Scorpion*, like her namesake, also carried a sting: a total of twenty-three torpedoes of three different types. Two of them had nuclear warheads. Her speed was rated at 20 knots on the surface and 35-plus knots submerged. (The navy may have kept her true speed secret to conceal it from hostile powers.) She was designed, like others of her class, to dive as deep as 2,000 feet and more.

On May 16, 1968, the *Scorpion* surfaced at the U.S. submarine base at Rota, Spain, and sent two crew members ashore on a tug. One had emergency leave to take care of a family problem, and the other left for medical reasons. On May 21, back at sea again, the *Scorpion* rose close to the surface, ran up her radio antenna to send a message to fleet headquarters, and slipped beneath the waves again. On May 22, at 6:44 P.M., oceanic listening stations picked up a single loud report followed by ninety-one seconds of silence and then a rapid sequence of crunching, tearing noises. On May 27, the *Scorpion* failed to appear at her U.S. base at Norfolk, Virginia. The navy declared her overdue. On June 2, she was declared missing.

The navy clamped a tight cover of secrecy over the loss of the *Scorpion*. "Mechanical failure," always a reasonable speculation and one capable of many interpretations, was given as the initial reason for her disappearance. Rumors, of course, circulated freely. One held that the *Scorpion* had been torpedoed by the Soviet subs she was trailing, either as a measure to preserve Soviet military secrets or as an act of revenge for the presumed sinking of a Soviet submarine by an American spy sub. Alternatively, some claimed that the *Scorpion* had been the victim of an accidental collision. The latter explanation was more probable, since neither side in the Cold War was anxious to initiate overt hostilities. It is said that the Americans and Soviets had an unspoken agreement not to raise the issue of submarine collisions, even when such an occurrence was fatal to one side or the other.

Yet another theory was that the *Scorpion* had struck an uncharted undersea mountain at 35 or 40 knots, going too fast for her sonar to detect the obstacle. And there were submariners who thought that the sub's main battery had exploded with catastrophic force, an event that had occurred before on other submarines. Or she might simply have strayed beneath her crush depth, as many submarines of many nations had before her.

Although the U.S. Navy probably knew very quickly where the *Scorpion* went down—some said within the hour—its releases said that it took six months to locate her. (Cynics whispered that the navy did not want to reveal just how effective its listening system was.) *Trieste II* was towed to the site and found the missing submarine in several pieces at more than 10,000 feet below sea level. The sail was torn off; the forward section lay in a trench dug by its own impact; the aft section, including the reactor compartment and the engine room, lay in its own trench. Observers on the *Trieste II* could see that the impact had been so violent that the engine room was telescoped into the reactor compartment. Around the major pieces of wreckage lay the usual small pieces of debris, such as insulation. Unlike its predecessor, *Trieste I,* the second *Trieste* was equipped with a manipulator arm and a basket, and it used these to bring back the *Scorpion*'s sextant, among other things. *Trieste* also took hundreds of pictures for a permanent, but highly classified, record of the wreck.

A Naval Court of Inquiry convened in the summer of 1968 and reviewed all the evidence: the photographs taken by *Trieste II,* the sound recordings, and the "paper trail" of messages sent back to Norfolk while the *Scorpion* was at sea, as well as letters written by crew to their families and mailed from ports. The seven high-ranking naval officers of the court considered over twenty scenarios. Three emerged as the most likely: the battery exploded, the hull imploded, or the vessel was struck by one of its own torpedoes. The last was finally deemed the most probable.

As this scenario ran, a torpedo with a high-explosive warhead (not nuclear) somehow became activated as it was being loaded into a firing tube (perhaps in a practice exercise). The crew decided to jettison it by firing it out the tube, a tactic they had used successfully with a dummy torpedo the previous year. This time, however, the torpedo, possibly fitted with a homing device, traveled in a circle and struck the *Scorpion,* destroying it.

The Naval Court completed its report in January 1969. The report was kept secret for twenty-four years, until 1993. Its release made a sensation in the press. The *Houston Post-Chronicle* published a series of articles in 1993 and 1995 that reflected very badly on the U.S. Navy. According to the newspaper, the navy had become dissatisfied with the amount of time that nuclear submarines spent in shipyards for maintenance and repairs—about 40 percent of their life. At the time, the navy's budget was severely strained by building new submarines to keep up with growing Soviet submarine power. At the same time, the navy was spending half a billion dollars to upgrade safety on submarines, spurred by the traumatic loss of the *Thresher.* Maintenance seemed a convenient place to cut expenses. Accordingly, the navy decided to sharply reduce maintenance schedules, and the *Scorpion* was selected as a test case. She was given "the cheapest and briefest submarine overhaul in Navy history."

The *Scorpion* was plagued with mechanical problems throughout her last voyage. Even one day after she left Norfolk, she began losing hydraulic fluid from her sail; the mechanism that controlled the forward diving planes was leaking at a rate of 50 gallons an hour. An estimated 1,500 gallons of the oily fluid went to waste. The captain complained that the hull was encrusted with barnacles, reducing the boat's speed by 1.5 knots. And some of the valves for pumping seawater out of the submarine were leaking. The leaking valves limited the *Scorpion* to a depth of 300 feet, although her rated limit was 2,000 feet.

A machinist wrote home on March 12 (the letter was mailed on one of *Scorpion*'s calls in port) that the crew had repaired, replaced, or jury-rigged every piece of equipment. Morale was poor because the crew felt that some of the officers disdained them and ignored their complaints about mechanical problems. None of these details, of course, was made public by the navy, which was zealous to guard its respected image.

So concerned was the navy about nailing down the actual cause of the accident, and thereby avoiding its repetition, that it sent the famous submersible *Alvin* with the tiny 3-foot ROV *Jason Junior* on a secret investigation of the site in 1986. *Alvin* served as a mother ship to *Jason Junior* and took video pictures of the little ROV flying around the wreckage. The results, however, were apparently inconclusive. The navy also monitored the site for radioactivity in 1968, 1975, and 1986 and reported that there had been virtually no leakage of radioactive materials from the ruined submarine's reactor or from its two nuclear-tipped torpedoes. Some cobalt 60 had leaked into the sediment from the reactor's cooling system, but its level had decreased over the years due to natural decay. No cobalt 60 was found in fish taken at the site or in undisturbed water samples. As for the plutonium, which might have leaked from the reactor, the navy reassured the world that its levels were no higher than what could be expected from fallout from various nuclear tests around the planet. (The navy did not mention the thousands of tons of nuclear waste dumped into the ocean by Russia, France, Britain, Japan, Belgium, the Netherlands, and other countries.)

* * *

From the *Thresher* to the *Kursk*, submarine technology has advanced by leaps and bounds, and the technology for locating, investigating, and sometimes recovering sunken submarines has kept up with the pace—if not exceeding it. Though the four tragedies discussed in this chapter cost millions of dollars and hundreds of lives, the forensic work accomplished by deep-sea detectives can often determine the causes of such disasters, giving closure to the families of the dead crew and officers and providing opportunities to remedy defects in the design of these remarkable underwater craft.

10 Developments in Detection: Outlook for the Future

It is difficult to predict just what the future will bring in the field of deep-water forensics. Hopefully, the world's major shipping nations will have learned from the analyses of past tragedies and taken steps to reduce the likelihood of new ones. But accidents will continue to happen as long as chance and human error play a role in seafaring, so there will still be a need for locating and investigating wrecks, particularly those that cause loss of life. Some of the most interesting developments in this area of deep-sea search and salvage pertain to improvements of sonar and advances in the design of autonomous underwater vehicles (AUVs).

Sonar is steadily becoming more refined, capable of transmitting increasingly clear images. One of the notable advances is the development of synthetic-aperture sonar, which combines the results of multiple sonar beams to obtain an image two orders of magnitude (100 times) clearer than conventional sonar. It can reportedly detect an object the size of a wallet at a distance of five-eighths of a mile.

There is also great excitement in the underwater exploration community about AUVs. First developed in the late 1980s, AUVs are true robots, operating independently according to the electronic instructions that have been programmed into them. A battery-powered "brain" sends these instructions to the torpedo-shaped craft's electric motors and steering devices. The battery also supplies power to the built-in sonar system and other electrical/electronic marvels.

An AUV can be launched from a ship or from shore—even, in a case of extremely high priority, from an aircraft. Once in the water, it gets its initial position from a global positioning satellite (GPS) and obediently follows its programmed instructions, going to a destination miles away, if necessary, and then "mowing the lawn" until it has gathered the data for which it was programmed. As it travels, it builds its own map so that it can find its way back to its starting point. In this context, a "map" can also be a sonar image of a wreck or a debris field.

To stay on course, an AUV may use an inertial guidance system, in which gyroscopes orient the vehicle in space and accelerometers sense changes in its speed and direction. This information is fed back to the central computer, which checks it against the programmed instructions and signals the electric motors and fins to make the appropriate corrections. Many AUVs use the less expensive dead-reckoning system, which maps the vehicle's course by constantly calculating its speed and direction, just as a ship's captain, in the days of sail, used to calculate his ship's position with data from the log, the compass, the chronometer, and the angle of the sun. The difference, of course, is that the AUV's guidance system works almost instantaneously and does not rely on hand calculations and mathematical tables.

As the AUV swims along, sonar sensors tell it how far it is above the sea floor, and the computer keeps it from deviating from the programmed height. A tightly focused sonar beam projected from the nose warns the AUV of obstacles ahead, such as a reef or a wreck, and the autonomous vehicle can steer itself around or over the obstacle.

AUVs were originally used for oceanographic research and military intelligence. Oceanographers were delighted to have this automated device that could map sea-floor topography, sample water temperature and salinity, measure the strength and direction of currents, and monitor the concentration of plankton. Already AUVs have examined the underside of the Arctic ice pack, and in the Antarctic they have monitored the concentrations of krill—tiny, shrimplike crustaceans that are a basic item in the oceanic food web. Scientists at Woods Hole Oceanographic Institution have employed them in conjunction with the famous manned submersible *Alvin,* using the AUV at night to locate sea-floor features such as hydrothermal vents while *Alvin* is having its batteries recharged and being readied for the next day's mission. This combination works because the AUV "sees" by the reflected sound waves of its sonar—it needs no light. The next day *Alvin* can go directly to the feature of interest and not waste time and battery power searching for it.

The U.S. Navy—and presumably other navies—make use of AUVs to locate mines and to map the sea floor. It is believed—though the navy likes to keep this sort of information classified—that AUVs were used in the recent war with Iraq. Other than locating mines, they can be used to search the floor of an enemy's harbor for wrecks and other underwater obstructions. Naval AUVs can also report on the location of ships of any nation, check sea-floor communications cables, and, in military planners' fantasies, be adapted to carry their own small torpedoes.

What does this have to do with deep-water forensics? According to a leading manufacturer of AUVs, economics is the key. Once the offshore oil industry (the major nonmilitary market for underwater work) begins using AUVs, the price will come down enough to make this technology practicable for locating and surveying wrecks. ROVs will still do the heavy work, but AUVs will be able to "mow the lawn" more cheaply and effectively than towed sonar and one day will be able to take images and photographs as well as ROVs.

There is a hitch, however, in using AUVs, and it is communication. While the ROV transmits its images directly to the surface through its tether and can be delicately manipulated through this cable, the AUV lacks this instantaneous communication channel. Researchers have been working on creating communication between AUVs and a controller on the surface, and already a human can transmit simple commands via sonar beam to an AUV performing its assigned task in the water. The AUV receives the signals through a sonar modem and can transmit sonar data back. At the time of this writing, such communication is too slow to be very useful, due to limited bandwidth. Only 9,600 bytes per second can be sent or received, which is not enough for maneuvering around a wreck. But as bandwidth improves, this impediment will be overcome.

"Hybrid" AUVs are currently used by the U.S. Navy and other organizations. These vehicles are connected to the surface by a fiber-optic cable that transmits signals back and forth almost instantaneously and gives the AUV great flexibility. The hybrid is not truly autonomous, however, and some models are designed to operate both autonomously and under control through a cable. AUV manufacturers point out that the hybrid still carries its own power supply. Whether or not such fine control is desirable depends on the mission: it is obviously not necessary for straight-line surveys, but it is very useful for inspecting a sunken vessel.

AUVs are manufactured in a whole range of sizes and with equipment varying according to the nature of their missions. Designing an AUV involves a series of trade-offs. The basic goal is to get the maximum amount of data gathering per dive, which translates into getting the most out of each battery charge. If the AUV works at a high speed, range is sacrificed; if maximum range is the goal, a lower speed is required (typical speed for a survey ROV is 3 to 4 knots). Speed is also problematical, since the resistance of the water increases as the cube of the velocity.

If the AUV needs to be able to reverse course and back out from a trap, such as an overhanging piece of wreckage or a sea-floor cliff, a reversing motor can be added but at a sacrifice of some other ability. An AUV can also be designed to hover over a wreck by adding thrusters. Equipped with this capability and cameras, it could take hundreds of still or video photos, from which a mosaic could be pieced together. In doing so, however, it would sacrifice forward speed (typical speed for a hovering AUV is 1.5 to 2 miles per hour). The Lockheed Corporation, using a prototype developed at MIT, has built a hover-capable AUV called CETUS. It is designed to locate mines but could be adapted for other purposes.

Battery power is, as it has been from the beginning, the limiting factor of AUVs, but some bold innovators are working to adapt fuel cells for use in these craft. This research may some day revolutionize the field. Fuel cells would be an ideal source of power for an AUV except for the compromises that such a power supply demands. Most fuel cells require a liquid electrolyte, which must be protected from contamination by seawater. But the pressure of the depths would make a pressure hull necessary to cradle the fuel cell safely. This pressure hull, like a miniature version of the pressure hull of a manned submersible, in which the pilot and others ride, would inevitably be massive and heavy. The added weight would be costly in terms of the extra power needed to run the AUV. And such a pressure hull would not be cheap. Most experts say that it will probably be years before a commercially practicable, reliable fuel cell for AUVs is ready. On the other hand, a Norwegian firm, Kongsberg Simrad, makes a fuel-cell-powered AUV called Hugin 3000, which is designed to dive to 9,840 feet and is currently used for inspecting offshore oil installations, pipelines, and communication cables in regions including the North Sea, the Gulf of Mexico, and off the coast of Nigeria.

The advantages of AUVs are several. For one, the AUV is not tethered to the surface like an ROV or a towed sonar. And the AUV can maintain a constant altitude over the seabed by means of its sensors and internal guidance system; an ROV or a towed sonar sled requires skilled piloting to maintain even an approximate altitude. The stability of the "platform" and the positioning of the sensors enable the AUV to send back images of astonishing clarity. Also, turnaround time is drastically reduced. To turn a towed sonar around at the end of a survey line takes hours while the mother ship slowly and carefully pulls the sonar sled in a huge circle at the end of hundreds or thousands of feet of tether cable. The AUV, in contrast, turns itself around and starts back on the next line in a matter of minutes. At the end of its run, it pops to the surface and broadcasts a radio signal of its position to the mother ship for easy recovery. Hoisted back on deck by the ship's crane, the AUV can have its data bank downloaded to the master computer on board while crew personnel take off the nose cone or removable panel and exchange the spent batteries for newly charged ones.

In addition to exciting technological advancements, there are other aspects of underwater forensics that must not be neglected in any attempts to predict the future. Locating a wreck and collecting evidence are only half the battle. To be useful, the evidence must be interpreted. According to a respected member of the Society of Naval Architects and Marine Engineers (SNAME), the debris field is the key to the story. Yet, as in the case of the *Arctic Rose*, this field is not always mapped. The trail of debris tells an expert whether the breakup occurred on the surface or as the vessel or aircraft sank and how rapidly it sank. The debris field was the key to unlocking the mystery of the space shuttle *Challenger* disaster in 1986.

However, fishing vessels are not covered by any laws that require mapping the debris field—mapping is governed only by voluntary guidelines. Although the sea where the *Arctic Rose* sank was not suitable for sport diving, in many other instances sport divers have plundered the debris field for trophies—and thieves have stolen items like propellers for resale. And so the evidence disappears. For this reason, SNAME hopes to persuade investigators of fishing-vessel sinkings to map the debris fields as soon as possible.

Another contribution made by careful analysis of debris is to ship design. Following the breakup of the tanker *Erika* off the coast of France in 2002, the European Union decreed that all tankers built in the future should have double hulls. This design introduces the risk, discussed in an earlier chapter, of corrosion in the empty space between the outer and inner hulls. As a countermeasure, it has been suggested that the space be filled with an inert gas or with plastic foam.

One of the goals of marine forensics, whether in deep or shallow water, is to reduce the chances of future accidents. Learning what caused a ship or a plane to go down enables bodies like the Coast Guard, the National Transportation Safety Board (NTSB), the Marine Accident Investigation Branch (MAIB), SNAME, and their counterparts in other countries to make this knowledge public with the hope that people will act on it. Stricter enforcement of safety regulations is a goal, as is the enacting of improved rules and regulations. Some of these guidelines seem painfully obvious, and others seem weak, but they are all steps in the right direction. A few of the recommendations are as follows:

- stability studies for fishing vessels under actual operating conditions, to be made available to the industry;
- tests with models to determine how much the International Load Line Convention needs to be amended with regard to require-

ments for hatch cover strength and freeboard for large bulk carriers;
- compulsory daily reporting of position by all vessels;
- electronic indicators to warn crews when hatch covers or ventilator caps are open or damaged;
- lighting and video cameras for the foredeck of giant bulk carriers like the *Derbyshire*, to enable the person on watch to see what is going on in a storm; and
- provision of survival suits for all personnel on board.

Such recommendations, *if enforced*, will greatly increase safety at sea. But there will always be accidents, and the occasional criminal sinking, so there will continue to be a need for deep-sea detectives to investigate the causes of these tragedies. Increasingly, the detection will be performed remotely, and perhaps remote techniques of investigation will be extended into relatively shallow waters. Finding the evidence will be carried out by ever more sophisticated machines, with less and less human intervention—but the evaluation and interpretation of that evidence will continue to be a job carried out by human deep-sea detectives.

GLOSSARY

American Bureau of Shipping (ABS) The classification society for ships built in the United States.

Atmospheric diving suit (ADS) A watertight, armored suit used for deep dives. In the suit, the diver breathes air at atmospheric pressure, which is the pressure of air on the surface. Technically, then, these suits are *monobaric.* (*See* JIM suit; Newt suit; SAM suit; WASP suit.)

Autonomous underwater vehicle (AUV) A true "robot submarine" that is self-guided and carries its own source of power.

Bends, the Also called *decompression sickness,* a condition involving pain, paralysis, and breathing difficulties due to the release of nitrogen gas bubbles in a diver's tissue. Occurs when air pressure is decreased too rapidly after a foray in a compressed atmosphere.

Bilge The lowest part of a ship's interior; also, the curved part of the hull between the sides and the bottom.

Black box The container that holds the cockpit voice recorder (CVR) or the flight data recorder (FDR) of an aircraft (usually, there is one black box for each recorder). Black boxes are conventionally colored bright orange for easy visibility but are called *black* because the data inside them are considered confidential.

Bulkhead A transverse partition or wall that separates one compartment from another on a ship; also, the partition between the pilot's cockpit and the passenger compartment on an airliner.

Classification society An agency that evaluates the condition of ships for safety. Without a classification society's approval, it is very difficult, and prohibitively expensive, to obtain insurance on a ship.

Cockpit voice recorder (CVR) A device in the cockpit of an aircraft that records the pilot's and co-pilot's verbal notations of flight events and conditions and their conversations; a verbal log of the flight.

Come-along A device for gripping and pulling a chain or wire cable, a sort of mechanical substitute for hauling in the cable or chain hand over hand.

Control van Located on board the mother ship, or support vessel, it is the enclosed space from which the pilots control, fly, and operate underwater search equipment such as ROVs.

Devil's claw A two-pronged device for securing an anchor chain and preventing it from running out and letting the anchor drop.

Doghouse A small structure on the foredeck of a big bulk carrier that covers an entryway to the ship's interior; used by crew members for convenience.

Dry dock An enclosure used for building or repairing ships, from which the water can be removed.

Emergency position indicating radio beacon (EPIRB) An important safety device containing a radio buoy. This device is kept on the deck of a ship; if the ship sinks to a predetermined depth (usually about 10 feet) below the surface, it is automatically released and begins broadcasting an emergency signal to guide rescuers.

Fathometer A type of echo sounder that measures the depth of water.

Flight data recorder (FDR) A device that records such data as an aircraft's airspeed, altitude, heading, and the like.

Frame In shipbuilding, the transverse structure that supports the hull plates and gives the hull its shape. In the days of wooden ships, it was called a rib.

Global positioning satellite (GPS) A navigation system that uses satellite signals to pinpoint the location of a radio receiver on or above the Earth's surface.

Green water A solid wall of water that crashes over a ship's deck; encountered in severe storms. The origin of the term is obscure, since "green water" is usually gray.

Gyrocompass A nonmagnetic compass that derives true north through a gyroscope (a wheel or disc mounted to spin on an axis oriented so as not to be affected by tilting).

Hatch An opening in the deck of a ship.

Hawser A thick rope or cable used to moor or tow a ship.

Heliox A helium-oxygen mixture used by divers at deep depths.

Hydrox A hydrogen-oxygen mixture used by divers at deep depths.

Hypothermia A dangerous medical condition caused by the lowering of body temperature.

JIM suit An ADS with jointed arms and legs and a depth capability of 2,000 feet or more. Mechanical pincers take the place of the diver's hands.

List The tilting of a ship to one side.

Magnetometer An instrument that measures the strength of a magnetic field. Towed by a helicopter or ship, a magnetometer can quickly locate a large iron or steel object such as the hull of a sunken ship.

Marine Accident Investigation Branch (MAIB) The organization responsible for investigating most marine accidents involving British vessels.

Marine Board of Investigation The United States Coast Guard agency that examines and evaluates evidence in marine accidents. A Marine Board of Investigation typically consists of three members.

Monobaric Having a pressure of one atmosphere.

"Mowing the lawn" A systematic back-and-forth pattern traveled by seacraft that allows seekers to collect as much data as possible about features on the ocean floor.

National Transportation Safety Board (NTSB) The organization responsible for investigating most marine accidents involving American vessels.

Newt suit An ADS named for its inventor, Philip Nuytten.

Nitrogen narcosis Also called *rapture of the depths,* a condition in which a diver becomes unsteady and disoriented. Occurs when nitrogen dissolved under pressure in the bloodstream affects the diver's mental processes.

Pinger A device that produces bursts of sound to mark or detect an underwater object.

Pitch The up-and-down motion of a ship from end to end; also, the angle of the propeller blades.

Remotely operated vehicle (ROV) A framework that carries the equipment necessary for underwater exploration. It is operated by pilots on board a support vessel on the surface. Often, but inaccurately, referred to as a "robot sub."

SAM suit An ADS with a depth capability of 1,000 feet. A SAM suit is usually smaller and lighter than a JIM suit, and prototypes were made completely of aluminum.

Screw Another term for the propeller of a ship.

SCUBA Also known as the *self-contained underwater breathing apparatus,* a device that employs a portable supply of compressed gas supplied to the diver at a regulated pressure for breathing underwater.

Society of Naval Architects and Marine Engineers (SNAME) An organization whose principal goals are to advance the art, science, and practice of naval architecture, shipbuilding, and marine engineering; encourage the exchange and recording of information; sponsor applied research; offer career guidance and supporting education; and enhance the professional status and integrity of its members.

Sonar An acoustic device that uses reflected sound waves to measure the depth of the water, locate sunken ships, and detect objects such as submarines and underwater obstacles. The term is derived from *sound navigation and ranging;* also known as *echolocation.*

Submersible A manned, nonmilitary submarine typically used for scientific research.

Tether A strong, flexible cable that connects an ROV to its support ship at the surface. It contains electrical wiring and fiber-optic cables as well as steel cables strong enough to support the weight of the ROV as it is lowered into or raised from the water. Also known as an *umbilical.*

Towfish A portable framework that carries sonar and is pulled behind the support ship.

Trim The fore-and-aft balance of a ship or an aircraft.

Trimix A mixture of three gases, such as oxygen, nitrogen, and helium, used by divers at considerable depths.

Turnbuckle A hollow, threaded device used to tighten (or loosen) metal rods or cables.

Umbilical *see* Tether

Voyage data recorder (VDR) A shipboard device analogous to the flight data recorder (FDR) of an aircraft.

WASP suit An ADS with movable arms but no legs; a cross between a submersible and a diving suit.

Yaw A type of ship motion in which the bow or stern moves rapidly from side to side.

SOURCES

Author's note: The original list of sources was considerably longer. Publishing considerations prevented it from being presented in its entirety. In part, this was because sources that the author downloaded from the World Wide Web were no longer available, since their URLs were no longer functioning; hence they could not be listed. Readers are cautioned that more URLs are liable to wink out after the time of this writing. An abbreviated list follows.

Note on abbreviations: In an effort to include as much material as possible within our spatial constraints, the following abbreviations have been used:

DETR	Department of the Environment, Transport and the Regions
HBOI	Harbor Branch Oceanographic Institution
ITF	International Transport Workers' Federation
MAIB	Marine Accident Investigation Branch
NTSB	National Transportation Safety Board
SNAME	Society of Naval Architects and Marine Engineers
WHOI	Woods Hole Oceanographic Institute

Prologue

Allen, Thomas B., "Remember the *Maine*?" *National Geographic*, February 1988, 92-111.

CHAPTER 1

Sleuthing the Depths: The Science of Underwater Forensics

PERSONAL

- Personal interviews with Richard Fisk and John Kreider of Oceaneering International, Inc., January 2001; William duBarry Thomas of SNAME, February 2001; Dr. Dana Yoerger of WHOI, December 1996 and June 2001; Dr. Robert Ballard, President of the Institute for Exploration, November 1996; and Donald Sheetz, Deputy Commissioner of Maritime Affairs, Republic of Vanuatu, December 2000 and January 2001.

PRINT

- Harris, Gary L. *Iron Suit*. Flagstaff, AZ: Best Publishing Co., 1994.
- Limburg, Peter R. and James Sweeney. *Vessels for Underwater Exploration*. New York, NY: Crown Publishers, Inc., 1973.

WORLD WIDE WEB

- Askew, Timothy, "History of Submersibles," HBOI, 2002, <http://www.hboi.edu/marineops/history.html>.
- Malone, Phil, n.d., <www.mr-phil.com> [Good general background on ROVs].
- Spawar Systems Center, San Diego, n.d., <http://www.nosc.mil.undersea>.
- U.S. Navy, n.d., <http://www.supsalv.org> [Site contains several links to information on ROVs. See also http://www.chinfo.navy.mil/navpalib/factfile/ships/ship_nr1.html].
- WHOI, n.d., <http://www.marine.whoi.edu/ships/rovs>.

CHAPTER 2

No Match for the Forces of Nature: The *Derbyshire*, the *Gaul*, and the *Flare*

PERSONAL

- Personal interviews and communications with Dr. Dana Yoerger and Dr. Jonathan Howard of WHOI, June 2001; Admiral John Lang (ret.), former head of the [British] MAIB, July 2000 and subsequently.

PRINT

- Broad, William, "New Tools Yield Clues to Disaster at Sea," *New York Times*, 16 March 1999, F1 *ff.*
- Clarke, Roger, "The Trawler *Gaul*: Why Was No Search Made for the Wreck?" London: DETR, 2000.
- "Found: The Wreck of the *Derbyshire*," *The Hull Daily Mail*, 7 June 1994.
- French, Andrew, "New Inquiry Demanded as Wreckage is Discovered," *The Newcastle Journal*, 8 June 1994.
- Guest, Andrew, "Probe Confirms Wreckage is *Derbyshire*," *Lloyd's List*, 9 June 1994.
- MAIB, "Marine Accident Report 4/99," Norwich, England: Her Majesty's Stationery Office, 1999.
- —. "Report of the Re-Opened Formal Investigation into the Loss of the *MV Derbyshire*," Norwich: Her Majesty's Stationery Office, 2000.
- —. "Report on the Underwater Survey of the Stern Trawler *Gaul* H.243 and the supporting Model Experiments August 1998 – January 1999," Norwich, England: Her Majesty's Stationery Office, 1999.
- Mulrenan, Jim, "'Structural Failure' hit *Derbyshire*," *Lloyd's List*, 7 July 1994.
- "Report of the ITF *Derbyshire* Mission," and various press releases, London: ITF, July 1994.

WORLD WIDE WEB

- Chief Inspector, "Report on the Underwater Survey of the Stern Trawler *Gaul* H.243 and the Supporting Model Experiments etc., Particulars of *Gaul* and the Incident," Chief Inspector's Script, n.d., <http://www.shipping.detr.gov.uk/maib/gaul/cis.htm>.
- Department for Transport Release, Britain, 25 June 2002, <http://www.dft.gov.uk/news.htm>.
- Derbyshire Families Association, n.d., <http://www.mv-derbyshire.org.uk>.
- "Holes Emerging in Trawler Spy Theory," BBC News, 11 August 1998, <http://news.bbc.co.uk>.

- Howitt, Wil, "Journal of Wil Howitt," n.d., <http://www.otolith.com/journal/whoi/>.
- *Hull Daily Mail*, 22 November 1997 – 11 May 2000; 25 July 2002 – 21 January 2003, <http://www.hulldailymail.co.uk/index.jsp> [Contact librarian if printouts are desired, as the online archive contains only the last six months of news].
- ITF, Maritime Department, Bulk Carrier Safety, n.d., <http://www.itf.org.uk>.
- MAIB, "Marine Accident Report 4/99. Report on the Underwater Survey of the Stern Trawler *Gaul* H.243 and the Supporting Model Experiments August 1998-January 1999," DETR, <http://www.shipping.detr.gov.uk/maib/gaul/synop.htm>.
- —. "The 2002 Survey of the *Gaul*," n.d., <http://www.fv-gaul.org.uk> .
- <www.ohmex.co.uk/gaul.htm>, n.d. [about the mini-ROV used to investigate inside the hull].
- <http://www.separation.org.uk/news/>, n.d. [deals with widows of *Gaul* crew].
- "The Unquiet Grave," n.d., <http://www.mediaworldnews.co.uk/page8.html>.
- Wainwright, Martin, "Human Remains Found on Trawler 28 Years After Mysterious Sinking," *Guardian Unlimited*, 26 July 2002, <http://www.guardian.co.uk/Archive/Article/0,4273,4468887,00.html>.

FILM/TELELVISION

- *Secrets of the* Gaul *Trawler*, Norwich, England: Anglia Television [Norman Fenton's film].
- *What Sank the* Derbyshire? Discovery Channel, 1998.

CHAPTER 3

The Ship that Toppled a Government: The Strange Case of the *Lucona*

PRINT

- Blum, Wolfgang, "Wie wurde die Lucona versenkt? (How Was the *Lucona* Sunk?)," *Die Zeit*, Germany, 22 September 1995.
- Brauner, Christian, "Ein Richter blickt per Kabel auf den Meeresgrund (A Judge Looks at the Sea Floor by Cable)," *Die Weltwoche*, Germany, 4 April 1991.
- Brunner, Erwin and Joachim Riedl, "Wien ahoi! (Vienna Ahoy!), *Die Zeit*, Germany, 31 May 1985.
- "Cake mit Sahne ([Yellow] Cake with Cream)," *Der Spiegel*, Germany, 24 April 1978.
- "Der smarte Herr Udo scheiterte schliesslich am Video (The Smart Mr. Udo Finally Ran Aground on Video)," *Die Weltwoche*, Germany, 14 March 1991.
- "Die Republik, die sich selbst verschenkte (The Republic That Gave Itself Away)," *Der Spiegel*, Germany, 5 December 1988.
- Mullen, Craig T., "The *Lucona* Affair," Eastport International, Inc. [now a part of Oceaneering International, Inc.], n.d.
- "Penetranz als Aufgabe (Mission: Penetration)," *Der Spiegel*, Germany, 16 June 1997.
- Polatschek, Klemens, "Hinrichtung eines Hofnarren (Execution of a Court Jester)," *Die Zeit*, Germany, 27 July 1990.
- *profil*, Vienna, Austria, 18 September 1989 – 11 March 1991.
- Santner, Inge, "Herr Udo, eine Uranmühle und sechs tote Matrosen (Mr. Udo, a Uranium Mill, and Six Dead Sailors)," *Die Weltwoche*, Germany, 29 October 1987.
- Steinhoff, Jürgen, "Ein Tausendsassa in Not (A Devil of a Fellow in Trouble)," *Stern*, Germany, 14 April 1988.
- "Voller Liebe (With Complete Love)," *Der Spiegel*, Germany, 5 February 1990.

World Wide Web

- Oceaneering International, Inc., "Investigation of the Sinking of the *M/V Lucona*," 1997-99, Oceaneering International, Inc., <http://www.oceaneering.com/oii.view/acc0003.htm>.

CHAPTER 4

Not-So-Benign Neglect: The *Marine Electric* and the *Rema*

Print

- Frump, Robert. *Until the Sea Shall Free Them*. New York: Doubleday, 2002.
- Geske, Robert, "VA Ship Firm Pleads Guilty to Felony in 1983 Sinking," *Pennsylvania Inquirer*, 21 October 1988, C19.
- MAIB, "Report of the Inspector's Inquiry into the loss of *MV Rema* with the loss of four lives on 25 April 1998 about 22 miles north-east of Whitby, North Yorkshire," Marine Accident Report 1/00, MAIB, Norwich: Her Majesty's Stationery Office, n.d.
- NTSB, "Marine Accident Report, United States Bulk Carrier *Marine Electric* Capsizing and Sinking about 30 Nautical Miles East of Chincoteague, Virginia, February 12, 1983," NTSB, NTSB/MAR-84/01, Washington, D.C.: United States Government, 20594.
- *Philadelphia Inquirer*, 13 February – 31 July 1983.
- U.S. Coast Guard, "U.S. Coast Guard Marine Casualty Report, S.S. *Marine Electric*, O.N. 245675. Capsizing and Sinking in the Atlantic Ocean on 12 February 1983 with Multiple Loss of Life," report No. 16732/0001, HQS 83.

World Wide Web

- Carr, Michael, "Winter Storms and Hypothermia," *Sail Net*, n.d., <http://www.sailnet.com/collections/articles/index.cfm?articleid=carrmi0040>.
- Search and Rescue Society of British Columbia (SARBC), "Hypothermia – Physiology, Signs, Symptoms and Treatment Considerations," SARBC, n.d., <http://www.sarbc.org/hypo1.html>.

CHAPTER 5

Unsolved Mysteries: The *Estonia* and the *Edmund Fitzgerald*

Personal

- Interviews and personal communications with Gregg Bemis, 3 November 2000, 2 December 2000, 16 February 2001, 1 October 2002, 4 November 2002.
- Personal communications with Anders Björkman, 11 October 2000, 24 October 2000, 25 November 2000, 28 November 2000, 25 January 2001, 26 January 2001, 15 June 2002; Tom Farnquist, Director, Great Lakes Shipwreck Historical Society, 28 March 2003 and 25 April 2003; Frederick Stonehouse, 31 March 2001; Ken Vrana, Director, Center for Maritime and Underwater Research Management, 19 November and 6 December 2002, 25 March 2003.

Print

- Bemis, Greg, and Jutta Rabe, "Report on the Diving Expedition in the Baltic Sea to the Wreck of the Passenger Ferry *Estonia* from August 19th to August 31st 2000," unpublished [can be downloaded (in Swedish) from http://www.kajen.com/~rasken/ms_estonia/ll.html], n.d.

- Bishop, Hugh E. *The Night the* FITZ *Went Down*. Duluth, MN: Lake Superior Port Cities, Inc., 2000.
- Björkman, Anders. *Nya Fakta om* Estonia*: Rapport of Estoniautredningen (New Facts about the* Estonia. *Report on the* Estonia *Investigation)*. Monaco: Editions EGC, 1998.
- —. *Lies and Truths about the* M/V Estonia *Accident*. Monaco: Editions EGC, 1998.
- Joint Accident Investigation Commission (JAIC) of Estonia, Finland, and Sweden, "Final Report on the Capsizing on 28 September 1994 in the Baltic Sea of the Ro-Ro Passenger Vessel *MV Estonia*," Helsinki, Finland: Edita Ltd., 1997.
- *Moderna Tider*, Sweden, December/January 2000; December/January 2001; October 2002.
- NTSB, "SS *Edmund Fitzgerald* Sinking in Lake Superior November 10, 1975," Marine Accident Report No. NTSB-MAR-78-3, NTSB, n.d. [download from <http://www.uscg.mil/hq/gcp/history/WEBSHIPWRECKS/EdmundFitzgeraldNTSB Report>].
- Stonehouse, Frederick. *The Wreck of the* Edmund Fitzgerald. Updated ed. Gwinn: Avery Color Studios, Inc., 1999.

World Wide Web

- Askew, Timothy M., "Submersible Surveys Wreck of the *Edmund Fitzgerald*," HBOI, n.d. <http://www.ch-somerville.com/bell.html>.
- Bishop, Hugh E., "*Edmund Fitzgerald*: 25 Years of Speculation, Fascination and Grieving," n.d. <http://www.lakesuperior.com/online/225/225fitz.html>.
- <www.CNN.com>.
- "Diving Expedition to the Wreck of the *Estonia* in August 2001," n.d., <http://www.balticstorm.com/english/diving_special_en.html>.
- The Independent Fact Group, n.d., <http://www.factgroup.nu.est/hole_00.html>.
- <http://www.kajen.com~rasken/ms_estonia>, n.d.
- Rönn, Cinna and Annika Sohlander, "Gregg Bemis och Jutta Rabe anhållna (Gregg Bemis and Jutta Rabe Arrested)," *Aftonbladet*, Stockholm, 9 October 2000, <http://www.aftonbladet.se>.
- Tarm, Eve and Michael Tarm, "The Sinking of the *MS Estonia*," *City Paper* (The Baltic States), Tallinn, Estonia; Riga, Latvia; Vilnius, Lithuania, n.d., <http://www.balticsww.com/>.
- Weidner, Mario M [leader of Bemis's and Rabe's dive team], "Operation *Estonia*," 2000, <http://www.kadel.cz/grid-tech/eng/en_estonia.htm>.

CHAPTER 6

Human Error or Just Bad Luck? The *Arctic Rose* and the *Margaretha Maria*

Personal

- Personal communications with Captain Ronald Morris, USCG, 8 March, 2002 and following; Lt. Commander Sue Workman, USCG, 2 March 2003; Richard Hansen, Marine Consultants of Puyallup, Washington, 12-13 May 2003; Dr. Bruce Johnson and Dr. John Womack of SNAME, 2002-2003.

PRINT

- Johnson, Bruce and Lt. George Borlase. *Time to Flood Analysis For the Fishing Vessel* Arctic Rose. U.S. Naval Academy and U.S. Coast Guard Marine Safety Center, 2003.
- MAIB, "Report of the Inspector's Inquiry into the Sinking of the Fishing Vessel *Margaretha Maria* BM148 with the loss of four crew between 11 and 17 November 1997," Marine Accident Report 3/99, MAIB, Norwich: Her Majesty's Stationery Office, 1999.
- Murphy, Kim, "Mystery of Boat's Sinking Explored," *Los Angeles Times*, 17 June 2001.
- *New York Times*, 3-5 April 2001.

WORLD WIDE WEB

- "*Arctic Rose* Sinks, Galaxy Burns, Who's Next?" n.d., <http://www.maritimeinjury.com/innernew.html>.
- "*Arctic Rose*'s Owner Chokes Up at Coast Guard Hearing," *North County Times*, 2001, <http://www.nctimes.com/news>.
- "Coast Guard Gets a Brief Glimpse of *Arctic Rose*," *The Olympian*, 19 July 2001, <http://www.news.theolympian.com>.
- "Coast Guard Releases Second Look at *Arctic Rose*," *North County Times*, 1 September 2001, <http://www.nctimes.com/news>.
- Injury at Sea—Maritime Lawyers [law firm], "Factory Trawler: No Time for Pain; No Time for Friends," n.d., <http://www.maritimeinjury.com/innerFACTORY.html> [excellent background piece].
- *Juneau Empire Online*, 13 June – 18 July 2001, <juneauempire.com>.
- Kenai Peninsula Online, 18 June – 12 July 2001, <http://peninsulaclarion.com/stories/061801/ala_061801ala0010001.shtml>.
- Klein Associates, Inc. [makers of Klein side-scan sonar], "The Shipwreck *Arctic Rose*," n.d., <http://www.kleinsonar.com/image/arctic_rose/arctic_rose.html>.
- "Letters from the Bering Sea," abcNEWS.com, 7 February 2002 [Search for story titles via <www.abcNEWS.go.com> homepage].
- "Lost with All Hands," abcNEWS.com, 13 February 2002.
- *Seattle Times*, 1 June – 3 August 2001, <seattletimes.com>.
- Murphy, Kim. "Fishing Where Few Even Dare," *Los Angeles Times*, 9 April 2001 <http://www.latimes.com/news/columns/colone/20010409/t000030611.html>.
- *Seattle Post-Intelligencer*, 20 June – 24 July 2001; 15 October 2003. <http://www.seattlepi.com>.
- *Seattle Times*, 18 – 31 July 2001; 14 October 2003, <www.seattletimes.com>.
- UFA Update, 22 June 2001, <http://www.ufa-fish.org/update/>.
- U.S. Coast Guard Marine Board of Investigation (MBI), Hearing Day 1, 9 July 2001, <http://www.uscg.mil/d17/ArcticRose/mbi01.htm>; 18 July 2001, <http://www.uscg.mil/d17/allnews/kodnews01/09601.htm>.

FILM/TELEVISION

- U.S. Coast Guard videotapes [untitled], 18 July 2001 and 24 August 2001.

CHAPTER 7

Intrepid, Independent Investigators: The Man Who Owns the *Lusitania* and His Peers

Personal

- Personal communications with Carole Bartholomeaux of Bartholomeaux/Public Relations, Inc., Agency of Record, Clive Cussler and National Underwater & Marine Agency (NUMA), carole@b_pr.com; Gregg Bemis; and Greg Stemm, December 2000 [see also the Web site of Odyssey Marine Exploration <http://www.shipwreck.net>].

Print

- Broad, William J., "Deep-Sea Clues to an Ancient Culture Discovered," *New York Times*, 12 October 1998.
- "Constitution and Bylaws of DEEPSEA," the Deep Shipwreck Explorers' Association [now Professional Shipwreck Explorers Association (ProSEA)], 1998.
- Cussler, Clive and Craig Dirgo. *The Sea Hunters*. New York: Pocket Books, 1997.
- —. *Dirk Pitt Revealed*. New York: Pocket Books, 1998.
- Johnston, Jo-Ann, "Tampa Salvors Land Ship Pact," *Tampa Tribune*, 6 October 2002.
- Stemm, Greg, "A Review of Deep Ocean Ship Exploration Issues—Part One: The Key to Davy Jones' Locker," *Ocean News & Technology*, March/April 1996.
- —. "A Review of Deep Ocean Ship Exploration Issues—Part Two: Pot Diggers or Adventurers? Historical Shipwreck Recovery Ethics," *Ocean News & Technology*, May/June 1996.
- —. "Cultural Heritage in Inner Space: Thoughts on the Future of Historical Shipwrecks," Greg Stemm for the Thirty-First Annual Law of the Sea Institute, 30 March 1998.
- Stemm, Greg and Captain J. Ashley Roach, "Shipwrecks in the Deep Freeze," *Maritime Heritage*, Volume 2, Issue 2, April 1998 and Volume 2, Issue 3, September–October 1998.

World Wide Web

- "These Are Some of My Favorite Things," Clive Cussler, NUMA Web site, n.d., <http://www.numa.net/> and <http://www.numaaust.nelsonbay.com/>.

CHAPTER 8

Tragedy in the Sky: Air and Space Disasters

Print

- Askew, Timothy M., "JOHNSON-SEA-LINK Submersibles' role in the CHALLENGER Recovery," Miscellaneous Publication # 24, n.d.
- HBOI, "JOHNSON-SEA-LINK I & II," Ft. Pierce, FL: HBOI, 2002.
- Langewiesche, William, "The Crash of EgyptAir 990," *Atlantic Monthly*, November 2001, pp. 41-52.
- *New York Times*, 3-19 September 1998; 1-14 November 1999.
- NTSB, "EgyptAir Flight 990, Boeing 767-366-ER, SU-GAP, 60 Miles South of Nantucket, Massachusetts, October 31, 1996," National Transportation Safety Board Aircraft Accident Brief PB2 002-919491, NTSB/AAB-02/01, DCA00MA006, NTSB,

n.d. [contains flight data and cockpit conversation between Captain Habashi and Captain al-Batouti].
- NSTB, "In-flight Breakup over the Atlantic Ocean Trans-World Airlines Flight 800 Near East Moriches, New York, July 17, 1998," National Transportation Safety Board Aircraft Accident Report PB2000-910403, NTSB/AAR-00/03, DCA96MA070, NSTB, n.d.
- Powell, Dennis E, "Obviously, a Major Malfunction," *Miami Herald Tropic* (Sunday magazine), 13 November 1988. [Also downloadable from <http://www.linuxandmain.com/features/challenger.html>.]
- "Report of the Presidential Commission on the Space Shuttle *Challenger* Accident." The Commission. Washington, D.C. 1986 (The Rogers Report). Selected portions follow [Note: items can also be accessed via links from http://history.nasa.gov.]: Bartholomew, Capt. Charles A. et al., "Appendix O: NASA Search, Recovery and Reconstruction Task Force Team Report," 8-14 May 1986; Beaton, John W. "Recovered External Tank Debris," and "Recovered External Tank Debris Hardware Assessment," Martin Marietta Corporation Launch Support Services, 10 April 1986; Feynman, F.P., "Feynman's Appendix to the Rogers Commission Report on the Space Shuttle Challenger Accident."

World Wide Web

- "A NASA Tragedy, Part 1: The Space Shuttle Challenger Disaster," n.d., <http://www.space.about.com/library/weekly/>.
- "A NASA Tragedy, Part 1: The Space Shuttle Challenger Aftermath," n.d., <http://www.space.about.com/library/weekly/>.
- Bresnahan, David M., "Bomb downed EgyptAir, say crash investigators," 5 November 1999, <http://www.greatdreams.com/bomb2.htm>.
- CBC News, 8 September 1998 – July 2002, www.cbc.ca.
- Fiorino, Frances, "TSB Sounds Alarm To Combat Inflight Fires," *Aviation Week*, 2002, <http://www.aviationnow.com/content/>.
- *Guardian Unlimited*, 21 November 1999 – 18 April 2001, <http://www.guardian.co.uk/egyptair>.
- Hailser, Mark A. and Robert E. Throop, "The *Challenger* Accident ... Mechanical Causes of the Challenger Accident and the Redesign Process that Followed," University of Texas at Austin: Mechanical Engineering Department, Spring 1997. <http://www.me.utexas.edu/~uer/challenger/challtoc.html>.
- *Halifax Herald*, 12 January – 16 April 1999, <http://www.herald.ns.ca/>.
- Jacinto, Leela, Margaret Litvin and Nasser Ibrahim, "Where is the Justice? Family of EgyptAir Copilot Enraged by Leaked Reports," 11 August 2000, <http://abcNEWS.go.com/>.
- Kagan, Daryn and Gary Tuchman, "Ships Track 'Ping' to Locate EgyptAir 'Black Boxes'," 2 November 1999, www.CNN.com [Item 96 on index under "EgyptAir 990"].
- Negroni, Christine, "Electrical Problems a New Focus in TWA Crash," www.CNN.com, 26 November 1996, <http://www.cnn.com/US/9611/26/twa.new.crash/index.html>.
- Scarry, Elaine, "Swissair 111, TWA 800, and Electromagnetic Interference," *New York Review of Books*, 21 September 2000, <http://www.nybooks.com/articles/13898>.
- Schultz, Gwyneth J.,"Accusations 'Reckless and a Mistake,' says Co-Author," U.S. Navy News Service, n.d., <http://www.chinfo.navy.mil/navpalib/>.

- Stoller, Gary, "U.S. Knew of Wiring Flaws Years Before TWA Crash," *USA Today*, 8 June 2001, <http://www.usatoday.com>.
- Transportation Safety Board of Canada, Communiqués, 21 October 1998 – 9 March 1999, <http://www.bst.gc.ca/en/media/emergency/communiques/swissair>.
- —. "Cockpit Voice Recorder CVR Recovery & Disassembly," 6 November 2002, <http://www.tsb.gc.ca/en/investigations/swissair/fac_updates/flight_recorder/cvr/cvr.asp>.
- —. Fact Sheets, 2 September 1998 – 27 March 2003, <http://www.tsb.gc.ca/en/media/fact_sheet>.
- U.S. Department of Defense, "Emergency Assistance – Disaster Assistance. USS *Grapple* to assist in the search of the Swissair Flight 111 near Halifax, Nova Scotia," *Defense LINK*, September 1998, <http://www.defenselink.mil/>.
- U.S. Navy, "Rescue and Salvage ships – ARS," U.S. Navy Fact File, n.d., <http://www.chinfo.navy.mil/navpalib/factfile/ships/ship-ars.html>.
- —. "Deep Drone 7200 Remotely Operated Vehicle," U.S. Navy Fact File, n.d., <http://www.chinfo.navy.mil/navpalib/factfile/oceansar/sar-ddrone.html>.
- —. "Mini Remotely Operated Vehicles," U.S. Navy Fact File, n.d., <http://www.chinfo.navy.mil/navpalib/factfile/oceansar/sarminirov.html>.
- —. "Navy Equipment Used in Underwater Salvage Operations," n.d., <http://www.chinfo.navy.mil/navypalib/oceanography/crash4.html>.
- —. "U.S. Navy Sends New Underwater Detection System to Halifax," 11 September 1998, <http://www.chinfo.navy.mil/navpalib/oceanography/swissair>.
- U.S. Navy Public Affairs Library, "NWSA1981. Navy's Role in 747 Search," Navy Office of Information, 25 July 1996, <http://www.chinfo.navy.mil/navpalib/news/navywire/nwsa96/nwsa0725.txt>.
- —. "TWA Flight 800—The Navy Divers," <http://www.chinfo.navy.mil/navpalib/oceanography/crash4.html> [Links to related topics].
- Williams Goddard, Ian, "Crash Damage Analysis" [A false but damaging TWV Flight 800 conspiratorial theory], n.d., <http://serendipity.nofadz.com>.

CHAPTER 9

Peril in the Silent Service: The *Kursk* and Other Submarine Tragedies

Personal

- Personal communications with Tom Dettweiler of Nauticos, 17 April 2003; Tim Janaitis of Phoenix International, 23 July 2002.

Print

- Baumann, Paul, "Ill-fated *Scorpion* Photographed," *New London (Connecticut) Day*, 17 September 1985, B1.
- Broad, William J., "Navy Says 2 Subs Pose No Hazards," *New York Times*, 7 November 1993.
- Drew, Christopher, "How *Scorpion* Killed Itself," *The Chicago Tribune*, 27 October 1983, 1.
- Komarow, Steve, "1960s Sub May Have Sunk Itself," *USA Today*, 27 October 1993, 1.
- "Navy Releases Submarine Files," Associated Press, 7 October 1983.
- *New York Times*, 17 August 2000 – 9 October 2001.

- Sheldon, Richard B. and John D. Michne, "Deep Sea Environmental Monitoring Conducted at the Site of the Nuclear-Powered Submarine Scorpion Sinking (R/V Atlantis II Voyage #117 Leg #1, August 19 – September 15, 1986)," Knolls Atomic Power Laboratory (Schenectady), KAPL-4749, October 1993.

World Wide Web

- Aeronautics.ru, "The Tragedy Aboard *Kursk*," 1 September 2000, <http://www.aeronautics.ru/nws002/kursk001.htm>.
- —. "What Happened to *Kursk*?" 25 January 2001 [An in-depth 43-page article].
- Beaver, Paul, "Russian Submarine Disaster—Latest," Jane's Information Group, August 2000, <http://www.janes.com/defence/naval_forces/news/jfs/jfs000815.shtml>.
- Bohmer, Nils, Frederic Hauge and Thomas Nilsen, "*Kursk*'s Potential Environmental Impact," *Bellona*, 16 August 2000, <http://www.bellona.no/data/dump/0/00/08/8.html>.
- Commander Submarine Force, U.S. Pacific Fleet, "USS *Scorpion* (SSN 589), May 27, 1968—99 Men Lost," n.d., <http://www.csp.navy.mil/othboats/589.htm>.
- —. U.S. Pacific Fleet, "USS *Thresher* (SSN 593) April 10, 1963—129 Men Lost," n.d., <http://www.csp.navy.mil/othboats/593.htm>.
- Conant, Eve, "Russia/Kursk Letter," *Voice of America*, 10 November 2000, <http://www.fas.org./news/russia/2002/001110.htm>.
- Cutting Edge, "Russians May Have Sunk the American Submarine, USS *Thresher* in 1963 Using HAARP-Type Particle Beam Weapons!" n.d., <http://www.cuttingedge.org/news/n1353.cfm>.
- dolphin.org, "INS *Dakar*," n.d., <http://www.dolphin.org.il/dakar> [more information at: "Search and Discovery of the Israeli Submarine *Dakar*," <http://www.dolphin.org.il/dakar/search/>].
- "Forensic Seismology Provides Clues to *Kursk* Disaster," American Geophysical Union, Release No. 01-5, 22 January 2001, <http://www.agu.org/sci_soc/prrl/prrl0105.html>.
- Goodenough, Stan and Patrick, "*Dakar* Discovered," focus-on-israel.org, 31 May 1999, <http://www.focus-on-israel.org/news88.htm>.
- *Guardian Unlimited*, London, 15 August 2000 – 9 October 2001, <http://www.guardian.co.uk>.
- "History of USS *Thresher* (SSN-593)," *Dictionary of American Naval Fighting Ships*, n.d., <http://www.history.navy.mil/danfs/t/thresher.htm>.
- Jane's Information Group, "Deep Submergence Rescue Vehicles," 14 August 2000, <http://www.janes.com/defence/naval_forces/news/jfs/jfs000814_3_n.shtml>.
- Johnson, Stephen, "Sub Sank in 1968 After Skimpy Last Overhaul," *Houston Chronicle*, 21 May 1995, A19, <http://www.txoilgas.com/589-news.html>.
- Kessel, Jerrold, AP, and Reuters. "Expert: Collision Probably Sank Israeli Submarine in '68," CNN.com, 1 June 1999, <http://www.cnn.com/WORLD/meast/9906/01/israel.sub/>.
- KurskSalvage.com, "The Salvage," 18 May 2001 – 12 November 2001, <http://www.mammoet.com/kursk/salvage/info.asp> [the best technical description of the salvage procedure, excellent pix].
- Kursk.Strana.Ru Network, 8 June 2001 – 10 September 2002, <http://www.kursk.strana.ru/english/dossier>.
- "Kyrgyzstan: *Kursk* Torpedo Not Factory Fault," *Washington Times*, 13 July 2002, <http://www.washingtontimes.com/upi.breaking/ ...>.

- Nassaer, Galaf, "Giving Chase to an 'Enemy' Sub," *Al-Ahram Weekly On-line*, Issue 401, 29 October – 4 November 1998, <http://weekly.ahram.org.eg/1998/401/eg2.htm>.
- NAUTICOS Corporation, "Maryland Company Locates Israeli Submarine," 1 June 1999, <http://www.nauticos.com/news/news/1999-06001.html>; "Maryland Company Salvages Israeli Submarine," 19 October 2000, <http://www.nauticos.com/news/news2000-10-16.html>; *Meridian Passages*, "Bringing Home the INS *DAKAR*," Volume 9, Issue 1, Winter 2002, <http://www.nauticos.com/news/2002_02_ newsletter01.html>.
- Nilsen, Thomas, Igor Kudrik, and Alexandr Nikitin, "The Russian Northern Fleet Nuclear Submarine Accidents," *Bellona*, 19 August 1996, <http://www.bellona.no/imaker?sub=1&id=11084>.
- Occupational Safety Observer, "A History Lesson: The Loss of the USS *Thresher*," June 1994, <http://www.disastercity.com/thresher/>.
- Offley, Ed, "Navy Says Sinking of the *Scorpion* was an Accident; Revelations Suggest a Darker Scenario," *Seattle Post-Intelligencer*, 21 May 1998, <http://www.seattlepi.nwsource.com/awards/scorpion/scorpion1.html>.
- RussiaLink, "K-141 *Kursk* Russian Submarine SSGN *Kursk* Catastrophe," n.d., <http://www.russialink.org.uk/kursk/events.htm>.
- Sontag, Deborah, "The Lost Sub Is Found, and Israelis Can Grieve," *New York Times*, 31 May 1999, <http://www.geocities.com/zincisrael/features/f62Lost_Sub_Dakar_Found.htm>.
- Subnet.com, "USS THRESHER (SSN-593)," n.d., <http://www.subnet.com/fleet/ssn593.htm>.
- SUBSIM Review, "*Thresher* Down," reprinted from *Mechanical Engineering Magazine*, February 1987 <http://www.subsim.com/ssr/thresher.html>.
- "THE *DAKAR*—Undersea Mystery is Finally Resolved," *Israel Today*, n.d., <http://www.israeltoday.co.il/article/Default.asp?CatID=14&ArticleID=52>.
- txoilgas.com, "USS SCORPION (SSN-589) Dedicated to the Memory of Our Fallen Shipmates," [posted by former crewmember Alan D. Stricklin], n.d., <http://www.txoilgas.com/589.html>.
- U.S. Navy, "U.S. Navy Ships—Wreck of USS *Thresher* (SSN-593)," Department of the Navy, Naval Historical Center, Online Library of Selected Images, n.d., <http://www.history.navy.mil> [follow links to Photographic Section—Naval Historical Center Web Site Search Engine; enter "USS *Thresher* (SSN 593)" in search box].
- —. "Wreck of USS *Scorpion* (SSN-589) Views Taken in 1968"; "Wreck of USS *Scorpion* (SSN-589) Views Taken after 1968," Department of the Navy—Naval Historical Center, Online Library of Selected Images, n.d., [links from] <http://www.history.navy.mil>.
- U.S. Navy Public Affairs Library, "U.S. Nuclear Powered Submarines Lost at Sea," 25 October 1993 [complete text of "white paper" released by the Department of the Navy], <http://www.ibiblio.org/pub/academic/history.marshal/military/navy/USN/sub_losses.txt>.
- Walsh, Don, "The Political Ocean," *Navy League: Sea Power Magazine*, Almanac 2000, 44-50, <http://www.navyleague.org/seapower//almanac/2000.htm>.
- Woodruff, Bob et al., "Message from the Watery Grave: Divers Recover Letter From *Kursk*," abcNEWS.com, 26 October 2000, <http://abcnews.go.com/>
- —. "Laid to Rest," abcNEWS.com, 2 November 2000, <http://abcnews.go.com/>.
- WPS Russian Media Monitoring Agency, <http://www.wps.ru/chitalka/en/index.shtml>, links to the following: Aleksin, Valery, "The *Kursk* Must Have Been

Rammed by a Foreign Submarine," *Nezavisimaya Gazeta*, 12-13 September 2000; Alinov, Victor, "In Masint's Cobweb," *Nezavisimaya Gazeta*, 7 September 2001, 10; Felgengauer, Pavel, "The *Kursk* was Killed by the 'Friendlies'," *Moskovskie Novosti*, No. 36, 4-10 September 2001, 7 [confirmation of the torpedo-explosion theory]; Kirillov, Yury and Ilya Bulavinov, "The *Kursk* Cannot Get Out of the Storm," *Kommersant*, 27 September 2001, 2; "Our Rescue Services Have Done Everything Possible and Impossible," *Krasnaya Zvezda*, 9 September 2002; Riskin, Andrei, Reactor in Shutdown Mode: Examination to Reveal More," *Nezavisimaya Gazeta*, 20 September 2001, 7; Saranov, Vadim, "Mutiny on the *Kursk*," *Versiya*, 10 April 2001, 3; Solovev, Vadim, "Dramatic Twist in the *Kursk* Disaster Investigation," *Nezavisimaya Gazeta*, 12 January 2001; Tokareva, Marina, "Everybody Knows Why The *Kursk* Went Down," *Obshchaya Gazeta*, No.3, January 2001, 1 [on the torpedo explosion]; "Underwater Cutting," *Izvestia*, 5 September 2001, 2, <http://www.wps.ru:8101/chitalka/kursk/en/09052.html>.

- Young, Michael, "Hazardous Duty: Nuclear Submarine Accidents," *Maritime Affairs*, 13 December 2000, <http://www.naval.ca/article/young/nuclearsubmarine accidents_by michaelyoung.html>.

CHAPTER 10

Developments in Detection: Outlook for the Future

Personal Communications

- Personal communications with Justin Manley of NOAA Ocean Exploration, 21 May 2003; Richard Dentzman of TRITON-ELICS, 22 May 2003; Peter Zentz of BENTHOS, 6 May 2003; Frank van Mierlo of BLUEFIN ROBOTICS CORP., 5 and 7 May 2003; Bill Kirkwood of Monterey Bay Aquarium Research Insitute (MBARI), 12 May 2003.

Print

- Clarke, Tom, "Robots in the Deep," *Nature*, vol. 421, 30 January 2003.
- Posey, Carl, "Robots of the Deep Blue Yonder," *Popular Science*, New York, February 2003.
- "Robot Submarines Go To War," *Popular Science*, New York, April 2003.

World Wide Web

- "Autosub Science Missions Thematic Programme," Autosub Science Missions NERC, Southampton Oceanography Centre, Southampton (UK), 2001, <http://www.soc.soton.ac.uk/PR/Autosub.html>.
- AUV Laboratory at MIT Sea Grant: "AUV Lab History: History of the Odyssey-class of Autonomous Underwater Vehicles," 16 April 2002, <http://auvlab.mit.edu/history.html>; "CETUS," 16 April 2002, <http://auvlab.mit.edu/vehicles>; "Caribou: Odyssey II Extended, Sonar Survey Autonomous Underwater Vehicle," 18 April 2002, <http://auvlab.mit.edu/vehicles>.
- Bluefin Robotics Corporation, n.d.:"Why AUVs?>Economic Drivers," <http://www.bluefinrobotics.com/auv.htm>; "Justifying use of AUVs," <http://www.bluefin robotics.com/reason.htm>; "Cost Savings," <http://www.bluefinrobotics.com/cost.htm>; "Autonomy," <http://www.bluefin robotics.com/autonomy.htm>; "Products," <http://www.bluefinrobotics.com/products.htm>.
- "Kongsberg Simrad AS – Sonar, Underwater Positioning and Survey Equipment," *Naval Technology*, n.d. [links from: <http://www.naval-technology.com/SPAWAR> Click on Products, then A-Z Company Index].

- "Side Scan Sonar," Klein Associates, Inc., n.d., <http://www.kleinsonar.com/discript/sssonar.html>.
- Doolittle, Daniel, "The Payoff Is in the Payload: Using AUVs in Scientific Research," *Underwater Magazine,* March/April 2003, <http://www.diveweb.com>.
- Jones, Daron, "The AUV Marketplace," *Underwater Magazine,* July–August 2002, <http://www.diveweb.com> [click on "Underwater Vehicles" and follow links].
- Morr, Brian, "All Quiet on the AUV Front," *Underwater Magazine,* January/February 2003, <http://www.diveweb.com>.
- Pilgrim, Greg, "AUV from Fugro, Oceaneering, and Boeing is a Reality," *Underwater Magazine,* May/June 2003, <http://www.diveweb.com>.
- "Remus AUV Plays Key Role in Iraq War," *Underwater Magazine,* July/August 2003, <http://www.diveweb.com>.
- Wernli, Robert, ed., "ROV Committee News," Vol. 6, Issue 3, Marine Technology Society, August 2002, <http://www.rov.org/news63.html>.
- "WHOI at Sea: Remote Environmental Monitoring UnitS (REMUS)," WHOI, n.d., <http://www.whoi.edu/home/marine/remus_main.html>.

ACKNOWLEDGMENTS

My thanks to Jim Lawrence and Mike Hanson of IMS, who were the keys to my first contacts; Don Sheetz, vice-commissioner of shipping for the Republic of Vanuatu; Gregg Bemis, for his generosity with his time and information on the *Lusitania* and the *Estonia*; Anders Björkman, for his invaluable help in the case of the *Estonia*; Philip Kimball, executive director of the Society of Naval Architects and Marine Engineers (SNAME), and Barry Thomas, the historian of SNAME, for their generous help in giving me a crash course in accidents at sea; Professor Alan Rowen, for his frequently solicited advice on technical points; Professors John Womack and Bruce Johnson, for their information on how conversion from one type to another affects the stability of fishing vessels; Shelley Dawicki, public information officer of the Woods Hole Oceanographic Institution (WHOI); Doctors Dana Yoerger and Jonathan Howarth of WHOI for their patient explanation of the investigation of the wreck of the *Derbyshire*; Richard Fisk and John Kreider of Oceaneering Inc., for the valuable background information they gave me on ROVs and for permitting me to tour their plant; Tom Dettweiler and Tim Janaitis of Phoenix International, for information on the *Dakar*; Bob Christ of VideoRay, for more background on ROVs and for an interesting tour of his plant; Adam Brown; Rear Admiral (ret.) John Lang, former director of the Marine Accident Investigation Bureau in the United Kingdom; Tamara Pleasant Crawford of the Freedom of Information Act office of the National Transportation Safety Board; the librarians at the Merchant Marine Academy at Kings Point, New York; my old friend Lars Landquist, former Swedish naval officer, for sending me additional material on the *Estonia*; Margy Walter for her cheerfully rendered help in translating German expressions in the case of the *Lucona*; Doctor Ludger Wess, for supplying me with articles on the *Lucona* case in the German (not the Austrian) press; and to all others who helped me in any way, great or small. I could not have written this book without them.

Oh, yes—thanks also to the people who put together Google, which was of inestimable aid as I trawled the Internet.